TRACTOR

The Heartland Innovation, Ground-Breaking Machines,
Midnight Schemes, Secret Garages, and
Farmyard Geniuses That Mechanized Agriculture

LEE KLANCHER

OCTANE
PRESS

Octane Press, First Edition
September 3, 2018
© 2018 by Lee Klancher

ISBN: 978-1-937747-95-4
LCCN: 2017964305

Cover and Interior Design by Tom Heffron
Project edited by Aki Neumann
Copyedited by Chelsey Clammer
Proofread by Dana Henricks

Printed in Canada

On the cover: The family of Landon and Lindsay Hunt (shown) have been farming this
land for three generations. The John Deere 4020 is owned by Jack Purinton
(shown in the seat) and is the second 4020 built and the first 4020 powershift. Lee Klancher

On the frontispiece: Starting a John Deere tractor near El Indio, Texas, in March 1939.
Russell Lee / Library of Congress LC-USF33-012101-M3

On the title page: 1917 Avery owned by Lou and J. T. Buice. Lee Klancher

On the copyright page: Steyr tractor at work in the mountains. Steyr

On the front endpaper: J. T. Buice and K. R. Hough at speed on a 1921 Rumely Oil Pull 30-60 Model "E". Lee Klancher

On the back endpaper: Australian workshop of Doug Rumboldt. Lee Klancher

On the back cover: International 3688 and J. I. Case Spirit of '76. Case IH; Fendt Vario 930. Fendt

octanepress.com

Contents

Introduction
The Machines That Changed the World

The tractor has changed the way the world operates, and is a vital piece of machinery that helped transform our world from an agrarian to an industrial society. In this book, you'll get a broad overview of the key machines that made transformation possible, as well as meet some of the people who engineered and designed these tractors.

The 161 tractors in this book were each selected for various reasons. Many of them are particularly innovative machines that represented a turning point in the arc of history. Others are interesting in what they didn't accomplish. The John Deere Model GP and International 560 are good examples. Both were machines rushed to market, and the flaws of each hurt both companies.

We've tried to include all the key brands in North America, with plenty of representation of overseas makers as well. Bear in mind that a book of this scope cannot include all 186 tractor makers from the 1920s, or every single brand worldwide. But you will find most colors have good coverage. A few tractors appear in this book simply because they are interesting, visually fascinating, or we found an image so incredible we simply had to include it.

We've also added a few things for those of you who have read all the books and know every detail. You'll find things in here that have not been published elsewhere, like the fascinating International TX-178 experimental and the concept drawings of the New Generation John Deere tractors done by Henry Dreyfuss Associates. There are also stories from behind the scenes, including industrial designer Chuck Pelly's entertaining account of his first design introduction gone horribly awry, and Bud Youle's clandestine work sneaking into the competition's private new model introduction.

I believe we've built a book that is visually engaging, and one that can be enjoyed by anyone interested in history, machines, or innovators. I hope you enjoy it, and would love to hear about your experience. Feel free to connect with us on our social media channels or drop us a note using the good folks from the postal service.

Lee Klancher

◀ **Nebraska collector Howard Raymond and his dog during a 2016 tractor ride to the Red Power Roundup in Des Moines, Iowa.**
Lee Klancher

Prologue
The Harshest Proving Ground

Prior to the invention of the plow and the reaper, farming was pretty much all Americans did. Ninety percent of the country's 3.9 million citizens were farmers. That meant building farm technology was not just a big business—it was arguably the biggest business.

As a comparison, in 2015 the US Census found that 80 percent of American households had a computer. Meaning, the potential market available to Google, Apple, and Microsoft was *a smaller percentage* of Americans than the markets entered by John Deere and Cyrus McCormick.

Interestingly, the two most important agricultural innovations would lead to the two largest makers of modern farm equipment. The descendants of Cyrus Hall McCormick, the man who invented the reaper, would found the International Harvester Company, which dominated the agricultural equipment market in the first half of the twentieth century and would eventually became Case IH. Likewise, descendants of John Deere, the inventor of the all-steel plow, would become, well, John Deere—the world's largest modern agricultural equipment manufacturer.

By the mid-1800s, the plow and reaper were mature technologies and the tractor was fermenting in the fertile fields of agricultural innovators. At that time, much of America had started to leave the farm, and in 1870, 53 percent of the American population were farmers.

As farming began to change steadily, Americans moved west to open new lands, even if doing so created a great personal risk. The country's population was exploding, and the 15 million farmers gave manufacturers a strong customer base of people who were literally dying for improved farm technology.

By this time, the use of steam on the farm grew significantly, driven by the need to power large threshing machines. Farm-oriented steam engines were offered on wheeled carts, which were typically drawn by horses.

Self-propelled steam engines were developed not long after, and would remain on the market into the 1920s.

In the 1890s, dozens of inventors and major manufacturers were experimenting with self-propelled, gasoline engine–powered farm machines that moved on wheels of steel and wood, as well as a few that ran on long corkscrew-embossed cylinders.

The race was on as farmers took new plots of land all around the world. Visionaries, former soap salesmen, powerful corporations, and backyard inventors built furiously in hopes of creating the machine that could replace the horse, improve the farm, and make the inventor rich and famous.

Chasing fame as an inventor was, and still is, a bit of a fool's errand, perhaps no more so than attempting to make a living working the earth. That seemingly bottomless American wellspring of hope drove farmers, dreamers, and traditionalists to plow their way through droughts, pestilence, and long odds to pursue a better way of life for their families and descendants.

◀ **Photographed in 1940, a young boy operates this machine under open skies in Grundy County, Iowa.** John Vachon / Library of Congress LC-USF33-T01-001821-M4

THE FIRST TRACTORS
1889–1906

The tractor was born in the late 1890s, as motive power developed. The early machines were raw, heavy, expensive, and appreciated only by the forward-thinking innovators and visionaries of their day.

1 · The Doomed Visionary

1892 Froelich Tractor · Origin: **Froelich, Iowa** · Company: **Waterloo Gasoline Traction Engine Company**

▲ The Froelich Tractor was the first successful farm tractor that could be driven in forward and reverse. The Froelich Tractor was constructed using a Van Duzen engine and a Robinson chassis.

LEE KLANCHER

The Driftless Area of eastern Iowa is a land of deep river valleys and bluffs that soar over the banks of the Mississippi River. The rolling countryside is covered in deep, rich loam, ideally suited for growing crops. Henry Froelich came from Kurhessen, Germany, to scout a location for his new farm, and found the region perfectly suited to his needs.

He bought land near Girard, Iowa, and brought over a group of twenty-eight immigrants in 1845, including his young son, John. One of nine children, John Froelich grew up ambitious and innovative. By the mid-1880s, he owned a grain elevator in the town of Froelich, Iowa.

Froelich had an inventor's soul. When he encountered problems, he developed new gadgets to solve them. Holding fourteen patents, Froelich invented a washing machine, a dishwasher, and an air conditioner. He also tried his hand at creating a corn picker, but when he took it out for testing and burned down the cornfield, he turned to other innovations.

Owning a steam engine and a J. I. Case threshing machine, Froelich took a crew to thresh grain in South Dakota on an annual basis. Prior to each trip, he built simple shacks that could be loaded onto rail cars or hayracks, and had them sent to South Dakota for his crew to live in during their months-long stay in the area. The shacks were also used by his young cooks to serve the crew's meals.

The shacks were essentially mobile homes—a concept that would not rise to popularity until the 1950s. "He may well have also invented the first mobile home," said Roger Swanson, a retired Froelich teacher and local historian.

Froelich was rightfully frustrated with aspects of the South Dakota threshing operation. Because the steam engine was fired up with shocks of corn, when the corn was damp in the morning or from rain, it was difficult to light. Considering the steam engines took about three hours to get up to power, they had to be fired before sunup to work a full day. The damp shocks caused delays and stoppages. And just to make things even more frustrating, when the shocks were dry, stacks of them could easily be lit by sparks from the steam engine exhaust.

Good water is also crucial to steam engines. A big one could use as much as 1,500 gallons in a day. If the water was alkaline, it could cause the engine to boil over or foam, thus spilling water into the cylinders. When that happened, work stopped until the cylinders were dried out. And even if the engine didn't foam, alkaline water had to be thoroughly rinsed out of the tank with clean water.

South Dakota had a lot of alkaline water. These issues with the water and fuel supply and the damp corn shocks all made Froelich lose time. He needed to find a better way to thresh his grain. When he purchased a gasoline engine for use on his farm in Iowa, he believed he had done just that.

In 1890, Froelich recruited a local man, William Mann, to help him create a self-powered gas engine to use on his operation in South Dakota. In a small blacksmith shop as well as the back of Froelich's general store, the two men worked to couple a single-cylinder Van Duzen engine to a Robinson chassis. The machine was capable of moving frontward and backward, and could power a thresher via a belt.

▼ **Inventor John Froelich developed his tractor to power a grain thresher. Froelich's tractor was innovative, but only two were sold.**
Lee Klancher

According to authors Margaret Corwin and Helen Hoy in their book, *Waterloo, a Pictorial History*, the machine refused to run the first time Mann and Froelich tested it. An empty rifle cartridge was then wedged into the priming cup to start the engine. Once running, the machine clunked along in forward and reverse.

In August 1892, the new self-powered machine and a crew of sixteen men were sent to South Dakota. The results were dramatic. Coupled to a J. I. Case thresher, Froelich reported that his machine harvested 72,000 bushels of grain in fifty-two days.

The crew's success inspired Froelich to work with investors and found the Waterloo Gasoline Traction Engine Company in

◀ **Several Froelich Tractor replicas were built, including this one housed at the store. The Froelich Tractor was the catalyst for the Waterloo Boy company, which was later purchased by Deere & Company.** LEE KLANCHER

1893. The company built and sold four tractors, two of which were returned by dissatisfied customers.

The company reorganized and focused on building more salable stationary gas engines, but Froelich lost interest and left the company. The Waterloo Gas Traction Engine Company would reorganize several times and eventually build the successful Waterloo Boy Tractor, which was purchased by John Deere in 1918.

After the failure of his tractor, Froelich, his wife, and their four kids moved to Dubuque, Iowa, after leaving the Waterloo Company. He worked for several engine manufacturing companies before moving to St. Paul, Minnesota, in 1929. There, he invented many other items, including a washing machine, which was dubbed the Froelich Neostyle Washer. He lived in St. Paul until his death on May 23, 1933.

Though his inventions did not result in fame or fortune, Froelich was inducted into the Iowa Inventors Hall of Fame in 1991.

▼ **John Froelich holds fourteen patents and invented a corn picker, washing machine, air conditioner, and a self-propelled reversible farm tractor.** US PATENT

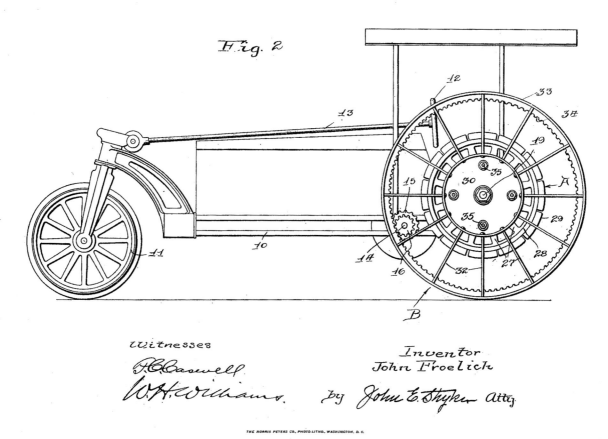

J. FROELICH.
COMBINATION TRUCK AND TRACTOR.
APPLICATION FILED MAY 10, 1915.

1,282,064.

Patented Oct. 22, 1918.
3 SHEETS—SHEET 2.

Fig. 2

Witnesses

Inventor
John Froelich

by John E. Stryker atty.

THE NORRIS PETERS CO., PHOTO-LITHO., WASHINGTON, D. C.

2 · The First "Tractor"

1903 Hart-Parr No. 1 · Origin: **Charles City, Iowa** · Company: **Hart-Parr Company**

▲ **Hart-Parr No. 1 was built in 1901 and is credited by the American Society of Agricultural Engineers (and many others) as the world's first internal combustion agricultural tractor.**

Shortly after Froelich introduced his tractor in Iowa, two young men met while registering for classes at the University of Wisconsin in September 1892. Both were engineering students, and they quickly discovered a mutual interest in mechanical devices.

The students, Charles H. Parr and Charles W. Hart, were a pair of young wunderkinds who, like Mark Zuckerberg and Steve Wozniak, married their affinities to opportunity and founded dynasties while in college. As students, Hart and Parr opened a small machine shop in Madison which they dubbed "Hart & Parr," where they repaired farm equipment and experimented with internal combustion engines. By the time the two graduated in 1896, they had more orders for tractors than they were able to build.

As soon as they graduated, the industrious pair founded a company, Hart & Parr, and built a two-story building on a half acre of land on Murray Street in Madison. In 1897, they sold some stock to raise capital and renamed their outfit the Hart-Parr Company, and recorded profitable business in 1899 and 1900.

By 1901, they realized they needed more room and were struggling to find both space and support in Madison. Mr. Hart grew up in Charles City, Iowa, and his father helped the pair obtain a good offer of a decent price on land and help with

▲ **Charles W. Hart and Charles H. Parr grew the Hart-Parr Company explosively, with annual revenue climbing from $454.33 in 1899 to $750,000.00 in 1908. Hart was the president early on, with Parr in a lesser role. Hart always made significantly more money than Parr as well.** Floyd County Historical Society

capital. The Hart-Parr Company moved to Charles City in 1901, and remained there until 1988.

Their first tractor, Number One (No. 1), was produced shortly after the move. During its 36-mile delivery to their first customer, David Jennings, the machine crashed through a wooden bridge into a muddy creek, sustaining only minor damage. The operator was unharmed, but complained that he lost his hat. The tractor was pulled out with a team of horses, repaired at the factory, and then delivered. It worked continuously for seventeen years after that and was scrapped just after World War I. The hat was never found.

The Hart-Parr Company is credited as having built the first successful internal combustion tractor. Its sales director, W. H. Williams, is falsely credited with coining the term "tractor" while writing advertising copy in 1906. The term had been used prior to that, notably in the patents of G. H. Edwards, who used it in 1880 to describe a crawler track design. In 1890, one of his patents described a self-propelled tracked machine he called a "tractor."

Regardless, the early success of Hart and Parr will never be doubted. The pair are credited as the makers of the first successful internal combustion tractor and have been dubbed "the founders of the industry."

THE TRACTOR PIONEERS

As steam and internal combustion power developed, the tractor was the logical next step in mechanizing the farm. Once it became efficient, powerful, and economical enough to replace horses, oxen, and mules, the tractor allowed a very few people to feed the entire world.

Charting out the trajectory of tractor development begins as one of those simple questions that gets more and more complex the deeper you dig into historical records. Broadly speaking, by the 1890s, a horde of backyard inventors, agricultural equipment manufacturer research and development teams, professors, and crackpots were dreaming about how the emerging technology of motive power could transform the farm.

The paths of the innovators were often complex, with many of them first building for small independents and moving to larger companies, either through acquisition or simply taking a new job once their startup went bust.

The period from 1900 to 1920 was somewhat like the computer boom of the 1970s, or the software boom of the 1990s, with many small players rising and falling and, after a time, fewer companies surviving and thriving in the marketplace (i.e., Apple and IBM in hardware, and Microsoft in software).

By 1938, the market would settle down into roughly a dozen major players and another few dozen specialty makers. At the first part of the twentieth century, however, a large cast of colorful innovators designed hundreds of machines ranging from the sublime to the ridiculous.

The technology was changing rapidly, and the rules for what constituted an effective tractor were being written, rewritten, and broken on a regular basis. Most of the major innovations in internal combustion and motive power were conceived between 1890 and 1920, and the agricultural machines from this time period are some of the most fascinating agricultural machines ever created.

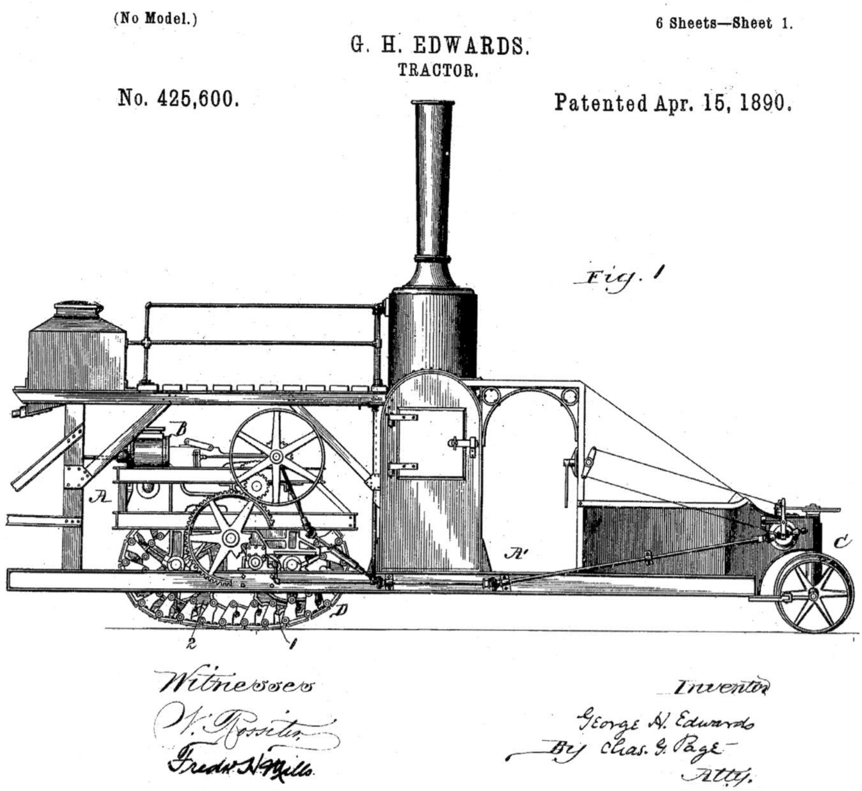

▶ **This 1890 patent is one of the first known recorded uses of the term "tractor" to describe an agricultural powered vehicle.**
US Patent

(No Model.)

6 Sheets—Sheet 1.

G. H. EDWARDS.
TRACTOR.

No. 425,600.

Patented Apr. 15, 1890.

Fig. 1

Witnesses

Inventor

George H. Edwards
By Chas. G. Page
Atty.

▲ The earliest form of motive power on the farm was provided by steam engines on carts. Case IH

▲ By the early 1890s, makers large and small experimented with internal combustion–powered machines, like this experimental Case tractor from 1892. J. I. Case

▶ Early designs of tractors were drawn by engineers who worked for large implement companies. This drawing was done by Deering Harvester Company engineers George H. Ellis and John F. Steward. Wisconsin Historical Society 115083

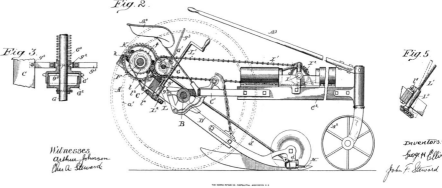

▶ As steam developed motive power, so did agricultural steam tractors. J. I. Case eventually offered massive steamers that had up to 150 horsepower. Case IH

THE PRAIRIE TRACTOR ERA
1907–1916

The earliest tractors were massive machines designed to plow unbroken prairie and power stationary equipment such as threshers. Like mainframe computers in the 1960s, these early machines were enormous, difficult to operate, prohibitively expensive, and made in very small numbers. CASE IH

3 · Old Reliable

1916 Hart-Parr Model 30-60 · Origin: **Charles City, Iowa** · Company: **Hart-Parr Company**

After introducing the No. 1, the Hart-Parr Company began selling internal combustion tractors sporadically. It was one of just a handful of companies selling these machines, and the pioneers found themselves battling the steam tractor industry's efforts to suppress its advertisements.

In 1905, Hart and Parr decided to go all in. They stopped the manufacture of stationary engines and committed wholeheartedly to building internal combustion tractors. The move proved wise. Despite expanding their factory and running the plant day and night, the growing company still wasn't able to keep up with the demand for its machines. By the close of 1906, the company had built and sold more than five hundred Hart-Parr tractors.

The next year, the Hart-Parr 30-60 was introduced. It would come to bear the moniker "Old Reliable," due to the maker's reputation for consistent performance. Expansion continued into 1909, and with the sales network growing across the globe, the company even briefly experimented with car manufacture, building at least two prototypes.

The company's growth was increasing so much that housing in Charles City became a severe problem. The company employed more people than the town could accommodate, and Hart-Parr built a hotel strictly to house and entertain its single male employees.

By 1910, Hart-Parr was the dominant maker of internal combustion tractors, with several sources crediting the company with selling more than 70 percent of the market. Whatever the number, Hart-Parr was the early market leader in internal combustion tractor manufacture, a position it would hold until a tidal wave of competition and questionable management changed its fortunes.

▶ **This Hart-Parr 30-60 was built in 1915. The twin-cylinder engine has a 10-inch bore and 12-inch stroke, and displaces 1,885 cubic inches.**
Buice Collection / Lee Klancher

4 · Steam to Gas to None

1921 Rumely Oil Pull Model "E" 30-60 · Origin: **LaPorte, Indiana** · Company: **Advance-Rumely Company**

▲ The Rumely Oil Pull Model "E" 30-60 was introduced in 1911 and produced until 1923. This one is a 1921 model.

Buice Collection / Lee Klancher

Advance-Rumely was one of the well-respected early agricultural equipment manufacturers. The company was founded in the 1850s by German immigrants Meinrad and John Rumely, who built and sold threshing machines and corn shellers. The company branched out and began manufacturing steam tractors in 1895.

In 1908, the company started developing internal combustion–powered tractors. The company brought in two talented engineers, John Secor and William Higgins, who had studied abroad with Rudolf Diesel. The pair pioneered an engine that would burn kerosene rather than gasoline. Kerosene was easier to produce and cheaper than gasoline, so the kerosene-powered engines proved popular at the time.

The company grew through acquisition in the early 1910s and folded the new companies into the Advance-Rumely brand. The company produced tractors (under the "Rumely" brand and also dubbed "Oil Pull") as well as engines, combine harvesters, and grain-processing equipment.

The Great Depression put the company up on the sales block, and Allis-Chalmers stepped up and purchased the company in 1931. The Rumely tractors were discontinued, but some of the harvesting equipment and dealerships lived on as part of the Allis-Chalmers organization.

▼ **The Rumely Model "E" was powered by a two-cylinder engine and weighed 26,000 pounds. This original condition example was used to build roads for POW camps in Oklahoma during World War II, before retiring to service as a collector's machine.** Buice Collection / Lee Klancher

5 · A Weighty Gamble

1913 Fairbanks Morse 15-25 · Origin: **Chicago, Illinois** · Company: **Fairbanks Morse**

▲ This 1913 Fairbanks Morse uses a single-cylinder engine and an open screen cooling system. The tractor spent its working life near the Gulf Coast in Texas, where the salt air severely deteriorated much of the machine. It's been restored to pristine condition.
Buice Collection / Lee Klancher

By 1910, internal combustion tractors were still new to the American farm, with US Census data recording only fifteen manufacturers building about four thousand machines. While this may sound like a fairly large number, bear in mind America had 6.3 million farms at that time.

One of the brands that jumped into the farm tractor fray happened to produce one of the world's best-known American products: the Fairbanks Platform Scale. That manufacturer became Fairbanks Morse, which built and sold windmills, gas and diesel engines, and beginning in 1910, farm tractors. Although various portions of the Fairbanks Morse company survived to modern times, its tractor production lasted only six years, with a handful of models built and production ending in 1915.

Fairbanks Morse left the tractor market just as it grew exponentially. In 1916, US Census data shows 114 tractor manufacturers producing 29,670 tractors. The race to build the machine that would transform agriculture attracted nearly every huckster, genius, agricultural company, and visionary in the world who had a hat in this very competitive ring.

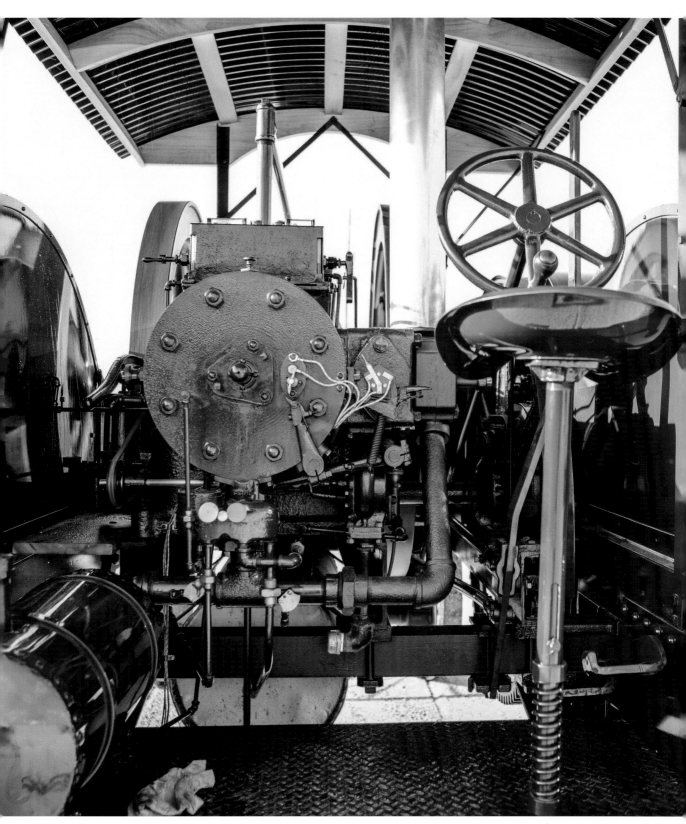

▲ The engine on the Fairbanks Morse is a single-cylinder with low-tension ignition. Buice Collection / Lee Klancher

THE FOUNDERS OF A GIANT

The story of one of the largest agricultural companies in history began in 1831 at Steele's Tavern in Virginia, where a young Cyrus Hall McCormick legendarily tested the first successful reaper, one of the most important innovations for the farm. It's possible that McCormick may not have been the first to do so—that's a hotly debated and expensively litigated issue—but the story has been made a legend mainly due to the fact that he built the most successful manufacturer of reapers, the McCormick Harvesting Machine Company.

McCormick began from scratch, then spent a decade struggling to obtain a patent and sell a handful of machines to often-skeptical farmers. After his father died, he and his brother Leander moved the company to Chicago, selected for its good access to water and rail transportation despite the fact that the city was rough-and-tumble (think Detroit today, plus organized crime, serial killers, and toxic pollution).

The company improved its product continuously, and proved to be brilliantly adept at marketing and organization; pioneering high-risk, high-reward sales methods, such as free equipment trials, money-back guarantees, and installment plans. It also held public, highly publicized field trials where different manufacturers' machines competed head-to-head. A modern-day equivalent might be Larry Page, the cofounder of Google, as he also used unconventional, high-risk methods to build a well-known brand around revolutionary technology.

Chicago and the McCormick's family situations grew concurrently. McCormick and his company prevailed despite court battles over patent rights, fires that destroyed factories and office buildings, and bare-knuckled salespeople brawling in the fields over equipment sales and competitions. By the end of the nineteenth century, the company emerged as one of the largest agricultural equipment manufacturers in America. It owned nearly all aspects of agricultural equipment production, and the headquarters was a 230-acre complex in the Chicago area known as "McCormick City."

Likewise, the McCormick family became one of the richest in the world and would rub elbows with (and marry, for that matter) members of other prominent American business families such as the Rockefellers.

Cyrus McCormick had a penchant for loudly voicing unpopular opinions, writing editorials for

▲ **The International Harvester Company (IHC) cautiously entered the tractor market in 1906, contracting with S. S. Morton, who supplied a friction-drive chassis that was mated to an IHC engine. This example was sold in 1907, and is pictured still at work in 1922.** Wisconsin Historical Society 70676

the *Chicago Tribune* and the *Daily Herald*. He was a devout Presbyterian, and donated significantly to a theological seminary in Chicago that to this day bears his name. McCormick believed his mission in life was to create technology that would help feed the world.

McCormick passed away in 1884, but his progressive, opinionated, and strong-willed wife, Nettie Fowler McCormick, and his descendants and relatives, would lead the company to continued success.

One the fiercest competitors of the McCormick operation was William Deering, the owner of a highly successful dry goods business who led his company, the Deering Manufacturing Company, into the agricultural implement industry in the 1870s by building innovative twine binders. He later turned to building harvesters and founded a plant in Chicago, Deering Harvester Works.

Deering retired in 1901, opening the door for change.

As the twentieth century dawned, a trend in American big business flared up. Giant mergers created companies that could dominate an industry. The Rockefeller family's legendary company, U.S. Steel, was perhaps the most notable of these mergers, but many other large corporations were also formed at this time.

This trend made its way to agriculture via J. P. Morgan, the legendary lawyer and financier who structured the U.S. Steel merger. A partner in the J. P. Morgan firm, George Perkins, brought the descendants of Cyrus Hall McCormick and William Deering to the table to discuss creating a new company. They agreed to merge, and subsequently swallowed up a number of other harvesting companies as part of the deal. In 1902, roughly 80 percent of the agricultural industry was merged into one large corporation, the International Harvester Company (IHC).

Right from its formation, this agricultural giant dominated the industry in the first half of the twentieth century. IHC's cash reserves, annual dividend payments, and sales network were far superior to any other agricultural company. The McCormick family assumed 43 percent of the initial stock, and the company expanded massively to own every single strand of the supply chain, from bolt manufacturers and steel companies for raw materials, to the railroads and ships that supplied IHC's vast network of factories.

McCormick family members remained in various leadership roles throughout the company's history, the last and best-known today being Brooks McCormick, who took the reins as CEO of IHC in the 1970s.

▲ **This 1912 experimental tractor is shown outside Deering Works. The label on the back of the photograph reads, "An early design of tractor built and never used."** Wisconsin Historical Society 24877

6 · The Hand-Carved Beast

1912 Titan 45 · Origin: **Milwaukee, Wisconsin** · Company: **International Harvester Company**

▲ The flagship of the IHC Milwaukee Works engineering group was this Titan 45, which was slightly shorter and more maneuverable than the Mogul built by Tractor Works in Chicago. Only a handful of these machines survives today.

Buice Collection / Lee Klancher

While Hart-Parr dominated the early tractor market, the giants of the farm industry watched and waited. Most of them entered the fray either with contract-built equipment or by acquiring tractor startup companies.

The largest of those giants was the International Harvester Company (IHC), which owned the agricultural equipment market in the early twentieth century. In 1910, they pursued the market vigorously with several different engineering groups developing machines. The Titan 45 was one of the largest and was developed by the engineers based in Milwaukee Works, the factory of the old Milwaukee Harvester Company that had been absorbed in the 1902 merger.

7 · Engineering Wars

1912 Mogul 45 · Origin: **Chicago, Illinois** · Company: **International Harvester Company**

By 1910, IHC had several engineering teams working to design tractors. The company had not yet properly integrated since the massive merger in 1902, and the legacy brands like Deering and McCormick built and sold tractors independently, often in the same markets or even the same city. The two strongest IHC tractor engineering departments were Milwaukee Works, which built the Titan line, and Tractor Works, which built the Moguls.

▼ This Mogul 45 was developed by the Tractor Works engineering team based in Chicago, Illinois. The design is distinctly different from the Titan also built by IHC on the opposite page.
MINNESOTA HISTORICAL SOCIETY

8 · Losing Steam

1916 J. I. Case Model 20-40 · Origin: **Racine, Wisconsin** · Company: **J. I. Case**

▲ The Case Model 20-40 (or Model 40) was introduced in 1912. This model ended up rusting away near a cotton gin in Lorena, Texas. The Buice family found it, rescued it, and then put it back into operating condition under a shade tree outside the Speegleville Store. BUICE COLLECTION / LEE KLANCHER

The J. I. Case Company is best known today for its powerful steam tractors. The company sold nearly thirty-six thousand steam traction machines between the 1850s and 1924, with peak production in the 1910s. J. I. Case dominated the industry worldwide, with no other company producing near that many steam engines. That said, the technology had significant drawbacks: steam engines were slow to fire, were heavy and cumbersome, required a licensed operator to run, and had the unfortunate tendency to explode catastrophically if improperly operated or maintained.

When steam tractor sales began to fall off in the 1910s, J. I. Case reacted by creating two large internal combustion tractors, the Model 60 (also known as the Model 30-60) and the Model 40 (also known as the Model 20-40), which were contract-built for them by the Minneapolis Steel and Machinery Company. It also built a new gas engine machine shop in Racine to construct the Model 40.

◀ The J. I. Case Company was founded by Jerome Increase Case in the 1840s, who grew it to become a leading producer of threshers and other agricultural equipment. Jerome I. Case was active in many roles in Racine, Wisconsin, including serving as the local bank president, supplying equipment to the local firefighters, and taking leadership roles in agricultural and scientific advocacy organizations. He also had a love for horse racing, steam ships, and auto racing. Case and his company owned one of the fastest schooners on the Great Lakes, the *J. I. Case*, owned a racehorse breeding farm in Kentucky, and even built a car to compete in the Indy 500. Case IH

▶ The *J. I. Case* was a 207-foot, three-masted schooner with a 190-foot mast. It was the largest sailing ship on the Great Lakes, and was also one of four large vessels jointly owned by Jerome Increase Case and Frank M. Knapp, a prominent Wisconsin business man. J. I. Case

THE FINE ART OF PRAIRIE TRACTOR RESTORATION

Dennis Powers's family began farming his place in Iowa in 1875, so perhaps it's fitting that the farm today—which is still being worked—is home for some of the earliest and rarest farm machines, the prairie tractors.

"I liked the oldest of the old, which started around nineteen-seven to nineteen-ten through the late twenties. The steel wheels, magnetos . . . no batteries, no upholstery," Powers said. He was interested in the larger tractors from the era. These monstrous machines plowed raw ground and powered stationary equipment.

The prairie tractors, as they are known today, were built and sold in limited numbers and were often purchased by collectives or co-ops who paid for the machines by traveling to different farms to work. The finish and engineering of the machines was state of the art for the time. Very few single farms could afford a prairie tractor, and very few of them survived into modern times.

Today, parts for these machines are rarely available. The only way to buy a part is to find another collector who happens to have one. When something breaks or needs to be replaced, restorers often turn to specialists to have the part made or expertly repaired. Many prairie tractor restorers have to find whatever records are available, have molds made, and cast parts. In short, they have to make parts to restore their machine.

In fact, prairie tractors are so rare and expensive today that only a few talented individuals are building tractors from scratch, using old literature and casting all the necessary parts. Others use computer numerical controlled (CNC) machines to make the parts they need. Specialists help all the restorers. Welders are some of the most sought after, and the good ones are artists.

"On a farm tractor, if you break a part on a cultivator, you weld it back on and it works. If you break it off a prairie tractor, you weld it back on and make it look like it never happened."

When the black clouds inherent to farming roll in, Powers can turn to his hobby for an escape. His extensive shop is a great place to work, or just to hang out and debate magneto timing and oiler maintenance. It's also a gathering place where his neighbors and friends love to stop in and just catch up.

▲ **The inside of the lean-to is a place to wrench as well as bench race. The two barber chairs are a favorite of most of the visitors. Haircuts, though, are not offered, nor are any other services. But owner Dennis Powers doesn't begrudge visitors a bit of shuteye. "I've even had people sleep in the chairs," Powers said.** Lee Klancher

▲ Prairie tractor restoration requires research and the ability to fabricate parts from blueprints or molds.
Lee Klancher

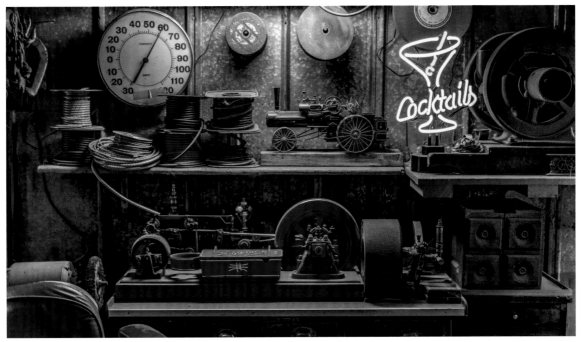

▲ The lean-to is a treasure trove for a collector. The model is of a Corliss steam engine. "The cocktails sign is a reminder for five o'clock," Powers said. "Not that we really need it." Lee Klancher

9 · The Triumph of Simplicity

1916 Waterloo Boy Model R · Origin: **Waterloo, Iowa** · Company: **Waterloo Gasoline Engine Company / John Deere**

▲ The Waterloo Boy Model R would evolve into the Model N, which was sold by Deere & Company after they purchased Waterloo Boy in 1918. The serial number of this Model R is 1460. Only about six machines exist with earlier numbers.

After inventor John Froelich left the Waterloo Gasoline Traction Engine Company, the organization was reorganized under new leadership, and it stopped building tractors and just built engines. Given they only sold two Froelich Tractors, this was hardly surprising.

In 1898, Louis Witry of Waterloo, Iowa, joined the company's twenty-person staff. Witry had been trained as a machinist, and had twelve years of experience working for locomotive companies. His industry expertise led him to quickly become the leader of the Waterloo Gasoline Engine Company, which flourished under his direction.

Meanwhile, just down the street in Waterloo, Iowa, the Associated Manufacturing Company had started development on a tractor, the Big Chief. When the company ran into financial difficulties, they sold off the Big Chief to the Waterloo Gasoline Engine Company.

Witry personally oversaw refinement of the Big Chief design tractor, doing most of the work in his shed in the backyard of his home at 303 Lafayette Street in Waterloo. That tractor became known as the Waterloo Boy, and was first sold on the market in 1912.

At roughly the same point in history, John Deere was selling machines from several other makers and experimenting with developing a number of its own, notably the Dain Tractor. The Dain was expensive to develop, and a complex, more technologically advanced machine.

John Deere management, concerned about the cost and timeline of the Dain, became enamored with the simplicity of the Waterloo Boy machine.

▲ **The Model R engine makes about 12 horsepower at the drawbar, and the machine weighs roughly 6,200 pounds. Top speed is about 2.5 miles per hour. The magneto ignition, gear drive, and fan-cooled radiator cooling system were state of the art in 1916.** Keller Collection / Lee Klancher

"I believe that quality and price considered, [the Waterloo Boy] is the best commercial tractor on the market today," wrote John Deere executive Frank Silloway. "The only real competitor it has is the IHC."

In 1918, John Deere bought the Waterloo Gasoline Engine Company for $2.35 million. The purchase came at a time when demand for tractors was accelerating. The Waterloo Boy tractors sold well in this environment, with John Deere moving more than nine thousand machines in 1918 alone. Witry stayed with John Deere as the factory superintendent at the Waterloo Boy plant.

The best part of the Waterloo purchase, however, was a design the company kept hidden until the last minute of the sale. That industrial secret was the raw beginning of a machine that would define John Deere farm tractors for more than forty years.

The drawing-board discovery would require more than six years of refinement and development before it could be unveiled. During this intense period of tractor innovation and growth, the face of John Deere's self-propelled farm machinery was represented by that simple machine refined in Louis Witry's backyard.

▶ **John Froelich continued to develop new ideas, even after leaving Waterloo Boy. This track system is similar to the Sure-Grip used at Waterloo Boy.** US Patent

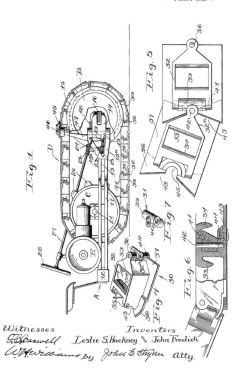

L. S. HACKNEY & J. FROELICH.
TRACTION VEHICLE.
APPLICATION FILED JUNE 24, 1914.

1,219,637.

Patented Mar. 20, 1917.
3 SHEETS—SHEET 1.

THE PLOW MAKER

John Deere was six feet tall, with broad shoulders and red hair. Born in Vermont, he spent his teen years as a blacksmith's apprentice and, at age twenty-one, started his own venture as a blacksmith. He tried various ventures, some with partners and others alone, none of which brought overwhelming success. Eleven years later, fires and the struggling local economy led him to declare bankruptcy. Not long after, in November 1836, he packed a few tools and headed to Grand Detour, Illinois, to seek better opportunities in the new portion of the country.

He was thirty-two years old and had to leave his four young children and pregnant wife behind. The legend goes that he had $73.73 in his pocket (about $1,900 in current value).

Deere built a shop and quickly found his blacksmithing skills in brisk demand. He began building and selling metal tools, including pitchforks that he would market by asking potential customers to step on the tines. They would spring back due to his heat treatment techniques, typically earning him a sale.

He then built a home, and Deere's family joined him in 1938. When they arrived, Deere saw his infant son, Charles Henry Deere, for the first time.

The thick matt of grasses covering the region's soil was of great interest at the time. The advantages of prairie soil were becoming well known, and opening the new lands was a difficult challenge. Back then, a popular way to do this was to hire a 125-pound prairie-breaking plow pulled with a team of up to

▲ This painting depicts John Deere and his first plow, which cut through sticky soil better than the heavy plow popular at the time. Thanks to Deere's sales and marketing, the chilled steel plow became a market leader. JOHN DEERE ARCHIVES

eight pairs of oxen. Other farmers preferred to break their own ground with smaller teams, and a lively debate emerged over the best method.

The soil in the region was legendarily sticky and thick, and would cling to the existing plows used at the time. Deere fashioned a plow from the smooth steel of discarded circular saw blade which cut cleanly through the soil.

Deere, a natural marketer, sold his new plow with vigor, and grew his business with illustrious integrity and independence. He briefly tried a partnership to grow the business and found it unsatisfactory. His dedication to quality was legendary. "I will never put my name on a product that does not have in it the best that is in me," he said.

In 1848, the company moved to Moline, Illinois, and incorporated in 1868. When his company grew to independence, Deere turned much of the duties of running it to his son, Charles. Deere then dedicated much of his time to civic duty, serving as president of the National Bank of Moline, director of the local library, and a trustee at the First Congregational Church, and served as the mayor of Moline. Deere died on May 17, 1886.

Very few American companies whose roots extend prior to the Civil War continue to survive today. Most have been merged, sold, consolidated, or dissolved. None of that applies to John Deere. The company's tradition of staunch integrity, careful consideration of when and how to enter a market, and intelligent management of resources and products begins with the company founder, John Deere.

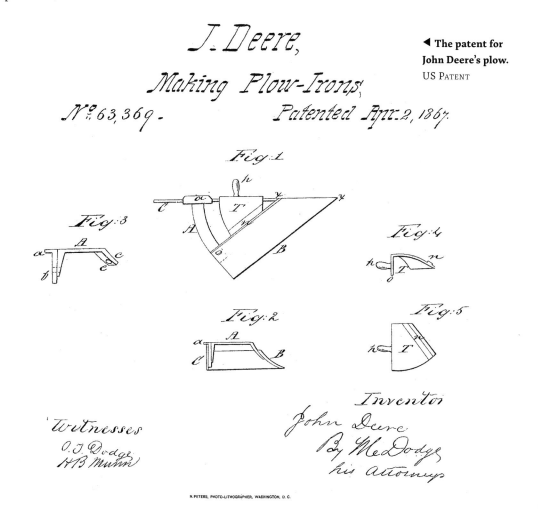

◀ **The patent for John Deere's plow.**
US Patent

THE FIRST WORLD WAR

By the 1910s, the world's largest exporters of grain were Russia, Canada, and the United States. As recounted in Ian Frasier's excellent book, *Great Plains*, a Turkish naval blockade stopped the flow of Russian grain to Europe in 1914, and the world was literally starving for grain. Demand for wheat increased dramatically and the price doubled in America, giving farmers increased revenue to purchase tractors. The new small tractors on the market were snapped up in great volume, and millions of acres of land were opened. The open land would play a role in the Great Depression, as the much of the open land would blow away in the Dust Bowl during the drought of the 1930s.

In addition to the wheat boom of the period, the war effort used tractors both to pull military equipment as well as to boost farmers' productivity, a much-needed assist as farms were struggling with labor shortages due to men being enlisted and dying in the war.

For better or worse, the mechanization of the farm can trace its roots to humanity's appetite for destruction, particularly when manifested in worldwide conflicts.

▲ **This image from December 1918 shows a 3-inch anti-aircraft gun mounted on a Caterpillar tractor.**
National Archives 111-SC-010508-ac

▲ **This January 1918 photograph of an Avery 40-80 being built in Peoria, Illinois, was taken as part of a series titled "The Tractor's Part in the War."** NATIONAL ARCHIVES 165-WW-318B-003

▲ During World War I, women were recruited and taught to work on farms. These Vassar College students were spending their summer vacation working on the school farm. NATIONAL ARCHIVES 165-WW-122B-021

▲ This small tractor is positioning a 207th squadron Royal Air Force bomber on August 18, 1918. The timing would make it part of the Second Battle of Somme. NATIONAL ARCHIVES / BRITISH WAR MUSEUM

▲ The United States employed more than four hundred thousand engineers to survey, build, and repair roads, bridges, and buildings during World War I. This group of engineers was being trained on how to use Avery tractors in January 1918.

NATIONAL ARCHIVES 165-WW-318A-031

▶ This tractor was tested on a 157-mile trip between Fort Sam Houston and Fort McIntosh in 1917.

NATIONAL ARCHIVES 165-WW 318A22

10 · Explosions Large and Small

1918 New Hart-Parr · Origin: **Charles City, Iowa** · Company: **Hart-Parr Company**

▲ The New Hart-Parr came to market in 1918. The "New Hart-Parr" moniker lasted only two years—in 1918 and 1919—and the gray paint is unique for the Hart-Parr line at that time. The tractor would evolve slightly to become the Hart-Parr 30. Dave Preuhs Collection / Lee Klancher

The Hart-Parr Company went through a hard time in 1914. It introduced the company's first small tractor, the two-cycle Little Red Devil, and it was a disaster. Overseas sales of all Hart-Parr tractors were terminated by the outbreak of World War I, and the domestic market responded to the war in Europe by collapsing to next to nothing as well.

Ever resilient, Hart-Parr focused on sales and got back on track in 1915, and in 1916, landed a $1.5 million military contract to produce shells for the British. The company sold $300,000 worth of tractors in the month of July *alone*, and ended the year with $4.5 million on their balance sheet.

This wave of success came crashing down again in the spring of 1917. In March, the company missed a key shell production deadline and the British canceled

▲ This tractor was built for the war effort and was found in the Library of Congress tagged as "Manufacturing War Materials" and dated Jan. 25, 1918. Library of Congress 32011

the $1.5 million contract, with no payments due. In April, America entered World War I, depressing sales. Then in May, Charles Hart was ousted from the company due to his penchant for racking up high levels of debt.

With classic midwestern industry and fortitude, new management came in and negotiated several lucrative military contracts and introduced a successful mid-size tractor, quickly righting the fiscal ship and increasing to more than one thousand employees.

After leaving the company he founded, Charles Hart took up ranching in Montana. The brilliant inventor had been a partner in the creation of the largest tractor company in early American history, and personally oversaw the creation of fourteen new tractor models.

11 · Long-Term Loan

1917 Avery 18-36 · Origin: **Peoria, Illinois** · Company: **Avery Company**

▼ **This Avery 18-36 was loaned to the agricultural engineering school at Texas A&M University. After the company went out of business, it was never recalled and stayed at the school for decades before being restored and owned by an alumni.**
BUICE COLLECTION / LEE KLANCHER

Founded while making steam engines, the Avery Company started building internal combustion tractors in 1910 with mixed results. After a company-designed single-cylinder performed poorly at the Winnipeg tractor trials, they contracted Albert O. Espe, a well-known freelance tractor engineer, to create a machine for them. The machine he designed was a mixed bag. The horizontally opposed four-cylinder engine in the tractor was quite good. The transmission was another story.

On the Avery, reverse was engaged by sliding the engine on a cradle. The design was less than satisfactory. The Avery Company also built trucks and crawlers, none of which provided enough sales for it to survive. The company struggled through the Great Depression, reorganized enough to make one last interesting tractor—the Ro-Trac, which could be converted from a wide- to a narrow-front setup—before closing its doors in 1941.

12 · The Name Game

1920 Titan 15-30 · Origin: **Chicago, Illinois** · Company: **International Harvester Company**

When much of the world's agricultural manufacturers came together under one roof to create International Harvester Company (IHC), they also amassed a large portion of the engineering talent. The group worked at a variety of facilities and on competing projects. The Titan and Mogul lines were marketed separately, and competing groups built them.

The company was deemed in violation of the Anti-Trust Act, and in 1914 had to sell off some portions and close competing dealerships in the same town. Most likely as a result of this, the tractors were branded "International" or "McCormick-Deering" rather than Titan and Mogul.

The now "unified" engineering team worked more or less in concert, or at the least under the thumb of engineering manager E. A. Johnston. The talented group worked furiously to take the lead in the hotly contested tractor market, and the 15-30 was an example of big budgets and expensive brains gone amuck.

The machine was known as Titan 15-30 in 1917, and later versions were dubbed the International 15-30. One of the key issues of the day was fuel economy, and the International 15-30 delivered this with water injection, and by starting on gasoline and switching to cheaper kerosene once running. To do this, the tractor had no fewer than seven carburetors, perhaps earning it the nickname the "Flaming Four."

In 1921, the Flaming Four was displaced by the much simpler and more effective McCormick-Deering 15-30, which in turn was overshadowed by a spindly looking thing known as the Farmall.

▼ **The International 15-30 was produced from 1918 to 1921, and is an evolved version of the Titan 15-30 built in 1917.**
Buice Collection / Lee Klancher

THE SMALL TRACTOR EMERGES
1917–1928

By 1917, the holy grail of farm equipment was the general purpose tractor. Dozens of companies and hundreds of innovators looked to build the machine that could replace the horse. While the entire industry struggled, one innovative American icon would use might and manufacturing to deliver the first blow, and another little-known inventor would create the mechanical innovation that would truly replace the horse.

13 · The Disrupter

1918 Fordson Model F · Origin: **Detroit, Michigan** · Company: **Henry Ford & Son**

▲ **This 1923 image shows the Fordson tractor, which featured a four-cylinder engine, three forward speeds, and no brakes.**
C. J. Hibbard / Minnesota Historical Society SA2.5 p78

Henry Ford grew up on a farm and developed a distaste for the drudgery of agricultural life in the nineteenth century. He left school at age sixteen, and took a job with Edison Illuminating Company. After a successful career and developing what would be a lifelong friendship with Thomas Edison, Ford founded the Ford Motor Company in 1903. Launching the Model T in 1908 successfully brought to life Ford's dream of affordable internal combustion transportation.

Ford's eye turned to the tractor in 1907, and he built a variety of experimental farm machines over the ensuing years. His experiments were fairly well known, with farmers holding out on purchasing machines and major manufacturers worrying about his entry into their market. John Deere's caution is well documented, in particular.

Due to a charlatan from Minneapolis trademarking the Ford name's use on tractors and introducing an inferior three-wheeled "Ford Tractor," Henry Ford was forced to start a separate company, Henry Ford & Son, for his tractor building endeavor. His tractors were known as "Fordson" until 1920, when he took back his corporate identity.

The Fordson tractor's first effort was a sale to the British government's Ministry of Munitions, who requested a demonstration. Their initial order was for six thousand Fordson tractors. The machine was introduced to the American market shortly after, with fewer than three hundred sold in 1917. In 1918, more than 29,500 Fordson tractors were sold, accounting for 22 percent of the US market. By 1928, Henry Ford had built and sold more than 675,000 Fordson tractors. His machines mechanized the farms of America and revolutionized the industry.

▲ The early Fordsons had a worm-gear final drive which caused the tractor to flip over backward when an implement encountered a rock or other immovable object. No matter—this model was the most successful farm tractor in history until 1924. PRINTS & PHOTOGRAPHS DIVISION / LIBRARY OF CONGRESS LC-F8-33725

14 · Overmatched

1918 International 8-16 · Origin: **Chicago, Illinois** · Company: **International Harvester Company**

▲ The International 8-16 was built at IHC Tractor Works and was powered by a four-cylinder engine mated to a three-speed transmission. A top speed of 4 miles per hour was quick for the era, and the tractor underwent three different evolutions over its production life from 1918 to 1922. Mecum Auctions 2018 Spring Classic / Lee Klancher

By 1917, small tractors were the sales leaders in the farm tractor industry, and the Fordson left almost everyone struggling to compete. The International 8-16 was a sleek, well-engineered entry into this market that should have set the agricultural world ablaze.

The International 8-16's sleek lines and layout were innovative, and the optional power take-off (authorized in 1919) was an industry first. Furthermore, the tractor was built on a crude chain-driven assembly line, likely one of the first to be manufactured this way at IHC.

None of this mattered much, as the Fordson was just as powerful, nearly as useful, and half the price.

15 · A Leap of Faith

1916 Case 9/18A · Origin: **Racine, Wisconsin** · Company: **J. I. Case**

An indication of how quickly the tractor market was shifting came when J. I. Case made a sudden move in 1914. The company was more than seventy years old, with eighty branch offices around the globe, multimillion-dollar annual revenues, and factories that covered sixty acres in Racine, Wisconsin, alone. Such entities generally don't make snap judgments, and J. I. Case had a reputation as a conservative company. In December 1914, company leadership authorized engineering leader David P. Davies to create a small, experimental tractor, powered by old stock 1912 four-cylinder auto engines. The three-wheeled contraption that resulted, the Model 10/20, was not a great machine with its poor weight balance, visibility, and traction. It was, however, introduced at the right time and sold well, with a total of 6,679 sold from 1915 to 1918.

In 1915, Davies was approved to build another small tractor, the 9/18A. This machine was a clean sheet design.

The Case 9/18A was a radical departure for the company; a small tractor with an all-new, kerosene-burning four-cylinder engine, and a lightweight body over a channel frame construction. The 9/18B replaced the A model and used a one-piece cast frame.

Although the model would only be produced from 1916 to 1918, the configuration would be used in the successful, well-engineered Case Crossmotor machines built for the next several decades.

▲ These women are being trained on tractor use during World War I. A Model 9/18 leads a Cletrac and an Avery tractor. National Archives 111-SC09869

THE GREAT TRACTOR WAR

Henry Ford and his Fordson changed the tractor landscape forever, and the battle between his company and the International Harvester Company (IHC) reshaped the tractor industry as well as the farm.

The times were not unlike the early days of the personal computer, with dozens of companies vying to create a new technology that would transform the world. Like the personal computer, the tractor promised to transform an industry, but early examples were too big and expensive for use on the average farm (or home).

If you think of IHC as Xerox or perhaps Hewlett-Packard—an established technology company with large resources and a relatively long history—and Fordson as Apple, you'll get a reasonable idea of the battle.

The wild card here is Henry Ford, who was like a supercharged Elon Musk. Like Musk, Henry had founded a company that captured the imagination of the public. Unlike Tesla—which has driven a niche market—the Ford car radically changed the automobile from a luxury item for the rich, to a meat-and-potatoes consumable for the masses.

After his success with the automobile, Ford became America's darling, his every move obsessed upon by the press. News that he would build an airplane was met with jubilation, and he was asked on several occasions to run for political office (he declined). Ford at one time started work on rewriting the Bible.

Ford had very deep corporate pockets and a zealous passion to bring power to the farm. He was willing to cut his margins to nearly nothing in order to sell his machines. The tractor industry had to respond to survive, and most manufacturers lowered prices in an attempt to gain some of the market that Ford had created. Tractor sales were up industry-wide, tractor

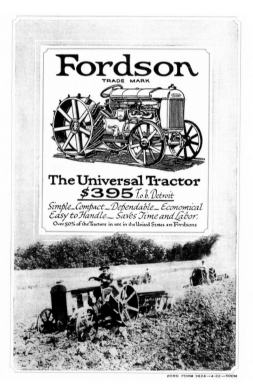

▲ **When America entered a rough economic stretch, Henry Ford sold his Fordson for $395—below his cost. This put the tractor on the farm in droves, and also drove the competition out of business.** Wisconsin Historical Society Press 96641

▲ **As Ford cut prices, so did International Harvester. In the tough market of 1921, IHC was able to hold on to market share, but when the Fordson price went down to $395, IHC lost share.** Wisconsin Historical Society

use on the farm grew dramatically over the next five years, and Ford's market share grew exponentially.

In 1921, in the midst of an economic and farm depression, IHC was fighting for sales. They dropped prices several times that year. They also initiated field demonstrations designed to showcase IHC tractors and convince buyers that IHC tractors were worth the extra money. The weapons were the Titan 10-20 and the International 8-16, and it seems they were up to the task of taking on the Ford. IHC regained a larger share of the sales that year, gaining 12 percent in market share. But Ford wasn't finished.

Ford responded in January 1922, dropping the price of the Fordson to $395, a price he admitted was below cost. IHC cut its prices as well, selling Titan 10-20s for $700 and International 8-16s for $670. The company also took the war to the fields, challenging Fordsons to plowing contests.

In *Century of the Reaper*, Cyrus McCormick had this to say about the tractor wars:

"A Harvester challenge rang through the land. Everywhere any single Ford sale was rumored, the Harvester dealer dared the Ford representative to a contest. No prizes were offered, no jury awarded merit to one or another contestant. No quarter was given and none was asked. Grimly the protagonists struggled, fiercely they battled for each sale. The reaper war was being refought with new weapons."

By 1926, IHC had its new McCormick-Deering tractors on the market and was regaining the dominance of the past. In 1928, Ford bowed out of the US tractor market, and IHC's new unit-construction McCormick-Deerings, plus the innovative Farmall, helped IHC take control of the market; 55 percent of all tractors sold that year bore the IHC logo.

Ford's timing was right, as the farm was ready for motive power. Also true is that the Fordson's affordable entry into the tractor market drove the market to grow.

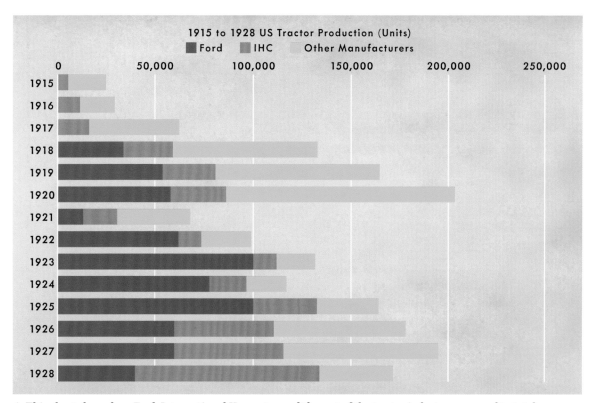

▲ This chart shows how Ford, International Harvester, and the rest of the tractor industry compared in total production and market share. Although the lowered prices and increased competition brought good things to the farmer, it devastated the industry, and IHC was no exception. Almost overnight, IHC went from controlling 39 percent of the market in 1916, to half of that in 1918 (see chart). US CENSUS DATA

16 · The Two-Cylinder Era

1923 Model D · Origin: **Moline, Illinois** · Company: **John Deere**

▼ The Model D is a shortened, highly refined version of the Waterloo Boy. Design advances included the unit frame, which used a cast tub to provide rigidity rather than the frame rails such as those on the Waterloo Boy.

Keller Collection / Lee Klancher

The architects of the Model D were steady-handed sorts, not the types to bend to the whims of a tractor-hungry audience. As dozens of opportunists rushed in to make farm tractors, Deere smartly limited their exposure by sticking with the tried and true Waterloo Boy Model N and Model R.

The machines were reliable but staid for the times. They were hardly a match for either Henry Ford's progressive Fordson or the more conservative machines like the Moguls and Titans built by IHC. Despite the fact that John Deere sold only seventy-nine tractors in 1921, management did not rush the Model D into production. In fact, management debated long and hard whether the company should even participate in the tractor market.

In the end, a refined Waterloo Boy was deemed an acceptable risk. Experimental models were built, tested, and refined.

So while Ford and IHC exchanged price cuts and traded profits for market share, John Deere quietly and economically developed the nicely balanced Model D.

Its power came from a refined version of the horizontal parallel twin-cylinder used in the Waterloo Boy Model N. Even in the early 1920s, a twin was a bit behind the times. Developing a four-cylinder engine was expensive, though, and John Deere decided to save money.

The John Deere commitment to two-cylinder engines would last until 1960, and the company's tradition of robust research and development of agricultural technology and cautious production would continue into modern times.

The first production Model D was built as a 1923 model, and it dramatically improved upon the Waterloo Boys. Although the unit frame and enclosed valvetrain soften and smooth the look, the hood is the key to the machine's handsome visage.

Good looks and a sturdy two-cylinder engine would serve the Model D well. Sales were solid, and John Deere kept refining the machine until 1953, making the D the longest-lived model in company history.

▲ **This 1926 Model D, painted orange at the factory, is considered one of the first machines the John Deere factory didn't paint green. Kay Brunner cast wheels were one of this tractor's distinctive features.**
KELLER COLLECTION /
LEE KLANCHER

THE TRACTOR MAKER BOOM ERA

When the Waterloo Gasoline Traction Engine Company was founded in the 1890s, a miniscule number of tractors were being built and sold by less than a dozen companies. That trend held steady through 1906 and 1907, the time when Hart-Parr rose to become the dominant maker of the very early days of the tractor. Into the 1910s, the agricultural industry realized that the tractor would be the next big development in farm technology, and the number of manufacturers grew steadily and then exploded.

In *Tractor History 1910–1934*, A. C. Seyfarth wrote that, "Overnight the farm gas tractor, which for years had been trying to break into the polite [s]ociety of approved farm machines, suddenly found itself basking in the sunshine of undreamed popularity."

US Census records show the growth in tractor industries peaking in 1921 with 186 manufacturers. The recession of 1921 also depressed sales, and manufacturers began dropping out quickly. By 1928,

fewer than a third of these companies remained. The Great Depression would weed out more. Only a handful of companies would survive to manufacture farm tractors in modern times.

▲ **Wallis tractors first appeared in 1912. Massey-Harris purchased the tractor line in 1928.** Wisconsin Historical Society 91920

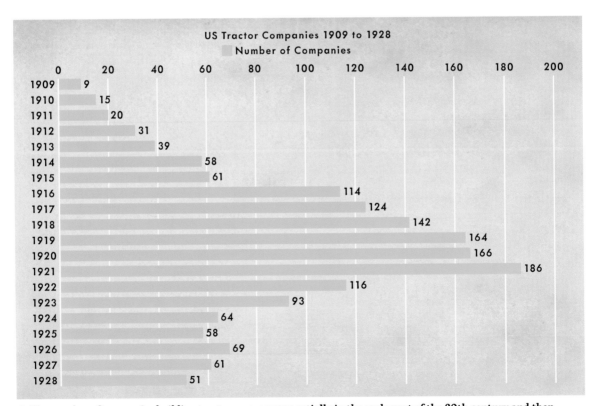

US Tractor Companies 1909 to 1928
Number of Companies

Year	Number of Companies
1909	9
1910	15
1911	20
1912	31
1913	39
1914	58
1915	61
1916	114
1917	124
1918	142
1919	164
1920	166
1921	186
1922	116
1923	93
1924	64
1925	58
1926	69
1927	61
1928	51

▲ **The number of companies building tractors grew exponentially in the early part of the 20th century and then dropped off.** US Census Data

▲ In 1919, the Vim Tractor Co. built a factory in Schleisingerville (later known as Slinger),Wisconsin. WISCONSIN HISTORICAL SOCIETY 90183

▲ Eagle Tractors were built from 1906 into the 1940s. The Eagle Manufacturing Company was based in Appleton, Wisconsin. WISCONSIN HISTORICAL SOCIETY 95714

▲ The Moline Plow Company was founded in the 1870s and purchased the Universal Tractor Company of Columbus, Ohio, in 1915. The Universal Tractor, shown here, resulted. WISCONSIN HISTORICAL SOCIETY 92860

◀ The Common Sense Gas Tractor Company was founded in Minneapolis in 1914 and sold the first V-8-powered tractor. The company was purchased by the Farm Power Sales Company in 1920, and production soon ended. WISCONSIN HISTORICAL SOCIETY 93706

▲ **The Gray Tractor offered terrific traction and an innovative cross-mounted motor. Built in Campbell, Minnesota, production lasted from 1916 to the late 1920s.** Wisconsin Historical Society 96714

◀ In 1925, the Holt Manufacturing Company and C. L. Best Tractor Company merged to found the Caterpillar Tractor Company. This is their Caterpillar 45 crawler tractor. WISCONSIN HISTORICAL SOCIETY 97448

▲ In 1918, the Trundaar Tractor was made by J. W. Lambert Buckeye Manufacturing Company in Anderson, Indiana. Very few were built, and only one is known to have survived. WISCONSIN HISTORICAL SOCIETY 97223

▲ The Bullock Creeping-Grip Tractor appeared in 1916, weighed about 7,200 pounds, and was out of production by 1922. WISCONSIN HISTORICAL SOCIETY 97236

17 · Singular Innovation

1924 Farmall Regular · Origin: **Chicago, Illinois** · Company: **International Harvester Company**

▼ The original Farmall was produced from 1924 to 1932, and later became known as the Farmall Regular. The tractor is credited as the true replacement for the horse due to its ability to plow, drive a belt, and cultivate mature crops.

<small>Dorothea Lange / Library of Congress LC-DIG-FSA-8B35395</small>

In the early twentieth century, agricultural innovators were obsessed with building a tractor that could replace the horse. The trickiest part of this was cultivation. A small tractor like a Fordson or a John Deere Model D could effectively plow a field and haul a wagon. The Achilles heel of such machines was that they did not have sufficient ground clearance to straddle (and cultivate) tall crops.

Bert R. Benjamin was one of the many brilliant engineers working on this issue. By 1915, he had spent twenty-two years overseeing the creation of IHC binders, shredders, and the power take-off (PTO) that debuted on the International 8-16 (credited as the first in the industry).

Not long after developing the PTO, Benjamin began to work on developing a motor cultivator. These were spindly machines with tall wheels designed exclusively for cultivation. Several companies built and sold them, and most were tippy, expensive, and ineffective. In 1919, IHC management directed Benjamin to cease and desist work on the machine.

Regardless, Benjamin persisted in developing the concept, with encouragement and occasional admonishment from IHC president Alexander Legge. In 1920, Benjamin had reversed the design and put the drive wheels in the back. The odd-looking machine was dubbed the "Farm-All." It was not popular with management, who were focused on the all-new McCormick-Deering 10-20 and 15-30, which were introduced in 1923 and expected to be the next great thing for IHC.

After much deliberation and lobbying by Benjamin, a very limited run of two hundred Farmalls was built in 1924, with many of them sent to Texas. The reaction by farmers was enthusiastic, prompting Houston IHC sales manager Jim Ryan to declare, "If you don't adopt [the Farmall] for production, we will organize a company in Houston and build it down here."

IHC management appeared to take the threat to heart, and by 1926 began to heavily promote it. Farmall sales went through the roof, the company regained the lead in the tractor market, and the Golden Age of IHC tractors began.

The Farmall was the result of one man's vision, while the McCormick-Deerings were designed and developed by a committee. Benjamin had an office at the swanky IHC corporate headquarters on Michigan Avenue in Chicago, but spent most of his time working on his designs out in an old shack next to McCormick Works.

Benjamin passed away at the age of ninety-nine in October 1969, honored with a raft of awards for his work. His Farmall lives on in history, credited as the first true general purpose tractor.

▼ **Bert R. Benjamin was a prolific IHC engineer, and the driving force behind the Farmall tractor. This patent shows one of his planter attachments for the Farmall.** US PATENT

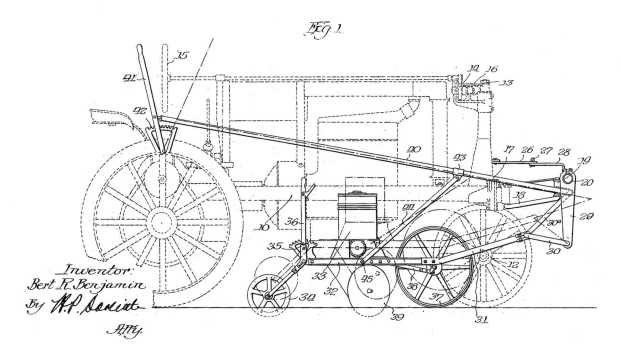

March 20, 1928.

B. R. BENJAMIN

1,663,236

PLANTER ATTACHMENT FOR TRACTORS

Filed July 30, 1926   2 Sheets-Sheet 1

Fig. 1

Inventor:
Bert R. Benjamin
By

ATTY.

18 · The Hart-Parr Evolution

1930 Hart-Parr 18-36 · Origin: Charles City, Iowa · Company: Hart-Parr

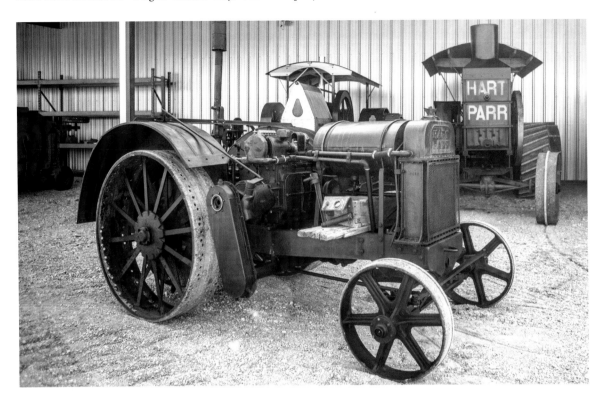

▲ The 18-36 was the second-to-last Hart-Parr tractor introduced before the merger with Oliver Chilled Plow Works. According to former Oliver chief engineer Herbert Morrell, some Hart-Parr 18-36s were equipped with a fully independent power take-off, making it an industry first. This 18-36 was purchased new by Lou Buice's grandfather, and the home-built modifications made by him remain.

Buice Collection / Lee Klancher

The Hart-Parr Company struggled in the early 1920s, battling both the depressed farm economy and the rise of the Fordson. They survived by producing stationary engines, electric washing machines, corn grinders, and other items that their factory was able to manufacture easily.

In 1925, they produced just over 1,000 tractors. That year, more than 160,000 tractors were manufactured, meaning Hart-Parr held less than 1 percent of the tractor market. The company balance sheet was strong, and by 1928, they upped production and were able to snare 3 percent of the market.

In 1929, the Hart-Parr Company merged with Oliver Chilled Plow Works, who made implements, to create the Oliver Farm Equipment Company. The new company acquired the American Seeding Machine Company, the McKenzie Potato Machinery Company, and the Nichols and Shephard Company, who made harvesters. The merger and acquisitions created a full-line farm equipment company and led to growth in the tractor business.

Notable is that the Oliver Chilled Plow Works was founded by Scottish immigrant James Oliver, who came to the United States in 1830. He ended up in Indiana, and in 1855 began making a steel plow, which became one of the leading plows of the day. The company grew and founded a plant in Hamilton, Ontario, that was later sold to IHC. Oliver also produced implements that were sold with the Fordson. In the mid-1920s, the Oliver Chilled Plow Works built over twenty experimental tractors.

After the 1929 merger, the new Oliver Farm Equipment Company closed its first year of existence by holding 7.6 percent of the tractor market, tied with J. I. Case, and behind only John Deere (who held 20 percent) and IHC (51 percent).

19 · The Quest for General

1928 Model C · Origin: **Moline, Illinois** · Company: **John Deere**

When John Deere needed a competitive general purpose machine, they turned to one of the company's star engineers, Theo Brown. The talented designer began the design on September 30, 1925. By October 9, he was sketching his idea on paper. The first production Model C experimentals appeared on March 15 of 1928. A few months later, Deere & Company issued a notice to their dealers that the name of the Model C was being changed to Model GP (for "general purpose").

The GP never became much of a sales success, but it is a beautiful piece of history.

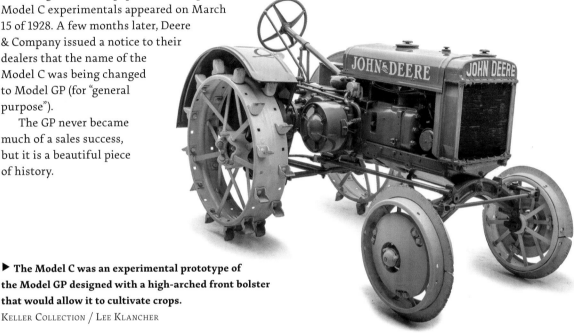

▶ The Model C was an experimental prototype of the Model GP designed with a high-arched front bolster that would allow it to cultivate crops.

Keller Collection / Lee Klancher

20 · Miles Ahead

1928 Massey-Harris GP 15/22 · Origin: **Toronto, Ontario** · Company: **Massey-Harris**

Massey-Harris was a large Canadian manufacturer of threshers and other agricultural equipment based in Toronto, Ontario. They began experimenting with tractors in the 1910s, selling Big Bull as well as Parrett tractors, some rebranded as Massey-Harris. In 1928, they purchased the Wallis tractor line and rebranded their entire Massey-Harris line. The company completed its own tractor, the GP 15/22, shortly after. The compact four-wheel drive was a brilliant idea, and technical execution was fairly sound. In tests, the machine didn't pull any better than comparable machines, such as the Fordson or John Deere Model D.

Massey-Harris would go on to bigger and better creations, but the GP 15/22 would end up a historical curiosity.

▲ The Massey-Harris GP 15/22 was a radical design with ordinary performance and was introduced in the Great Depression. Not surprisingly, sales were slow, with only about one thousand produced. Super T © 2017

TRACTORED OUT

When the stock market crashed on October 29, 1929, it triggered the worst economic downturn in the history of the industrialized world. By 1933, 15 million Americans were unemployed and half of the banks had failed.

This cataclysm had a disastrous effect on farmers. When the harshest tariffs in a century were enacted in 1930—the Smoot-Hawley Tariff—foreign markets retaliated with their own tariffs, depressing crop prices so low that it was more economical to burn corn for heat rather than sell it.

Farmers banded together in unions, blockaded milk deliveries on country roads, formed the occasional angry mob that threatened to lynch the local bank president, and stonewalled foreclosure auctions.

None of that made much of a difference.

From 1930 to 1935, some 750,000 farms were shuttered.

The tractor industry sold very few machines from 1930 to 1933, and the poorly financed or managed companies went out of business.

The tractor was a new technology, and with it came both positive and negative change. Although it allowed fewer people to feed the world, playing a not insignificant role in the modernization of our world, the tractor also displaced large numbers of farm laborers and tenant farmers.

The term for a farm worker being replaced by a tractor was "tractored out." That happened organically as tractors appeared, then accelerated in the Great Depression, and occurred even more frequently during the New Deal, when government funding was made available to farmland owners to fund mechanization.

In response, the Farm Security Administration was set up in 1937 to help displaced farmers.

▶ **This seventy-year-old tenant farmer was photographed near Goodlett, Texas. He and his sons were "tractored out," meaning the landowner replaced them with tractors. This practice happened regularly in the 1930s.**
Dorothea Lange /
Library of Congress
LC-USF34- 018175-E
[P&P] LOT 547

▲ This image shows R. Germeroth, a farmer in Sheridan County, Kansas, who purchased his John Deere tractor with a Farm Security Administration loan. RUSSELL LEE / LIBRARY OF CONGRESS LC-USF33-012358-M4

THE RISE OF STYLE
1929–1944

As the farm economy began to recover from the Great Depression, world-famous industrial designers and buttoned-down slide rule engineers worked in concert to add color and style to agricultural machinery.
MINNESOTA HISTORICAL SOCIETY, JENSEN 827

21 · The Color Revolution

1929 Allis-Chalmers Model U · Origin: **West Allis, Wisconsin** · Company: **Allis-Chalmers Manufacturing Company**

By the the 1930s, tractors were simple enough for an average human to operate and prices were such that farmers with larger operations could afford them. The tractor, however, was not yet a standard staple on the farm, with one tractor for every thirty working farmers. The Depression had reduced the number of farmers and devastated the survivors' cash reserves.

Tractor technology was mature and farmers needed that new tech to work ever-increasing acreage. The trick at the time was to get farmers to part with their hard-earned cash for a new machine.

Marketing was undergoing a revolution at the time. Agricultural equipment companies were already quite sophisticated when it came to marketing and selling, and were some of the pioneers of selling on credit and offering installment payment plans. The industry also used sophisticated and attractive advertising, in-field demonstrations and competitions, and aggressive, savvy salespeople who traveled to (and understood) customers.

One of the revolutions taking place at the time was the understanding that how a machine looked was nearly as important as how well it functioned, and sharp leaders were quick to adopt these new tricks to help get farmers into the seats of their new equipment.

▼ This 1929 Allis-Chalmers Model U was photographed at the Davenport Speedway in Iowa. MECUM AUCTIONS / LEE KLANCHER

One of those pioneering companies was Allis-Chalmers, whose roots went back to Edward P. Allis and the bankrupt flour mill equipment manufacturer he bought at a sheriff's auction in 1863. The company grew through mergers and acquisitions to offer a wide array of goods and services, primarily for the mining industry. In 1912, they began building internal combustion farm tractors and implements.

In 1926, the company hired Harry Merritt to help bolster the tractor business. Merritt, who had a long career at Holt, expanded their tractor business, and one of his key innovations for the entire industry came about when he drove past a bright orange field of poppies in California. The "powerful attraction" of the colorful flowers legendarily inspired him to paint Allis-Chalmers tractors Persian Orange. The splash he created resulted in the entire market embracing bright colors, and tractors soon became known by their color as much as their brand.

The Allis-Chalmers Model U earned fame not for features or color, but for tires. Ever the innovator, Merritt worked with his friend Harvey Firestone to put rubber tires on tractors. Early testing showed that when run at 12 pounds per square inch, which was a relatively low pressure for the times, rubber tires offered vast performance improvements over steel wheels.

Allis-Chalmers offered the tires on the Model U and promoted it with a cross-country tour of the machine. Driven on local racetracks by famous race car legends such as Barney Oldfield and Frank Brisko, the promotion highlighted the Model U's tires and high-speed fourth gear.

The appeal of speed was an effective marketing tool that worked to promote the tires, which further resulted in the farming industry adopting one of the most critical technological advances.

The use of Persian Orange was also a tremendous success, and kick-started the era of visual innovation transforming raw technology into functional pieces of art.

"U" FOR UNITED

The Model U shown here is significant for not only its color, but also why it was created. The machine is similar to a Fordson, and this was no coincidence. When the Fordson was pulled from US distribution in 1928, dozens of companies who sold implements and other equipment were facing the loss of a major revenue source. Many of these companies banded together to form the United Tractor Company. The company was designed to offer a Fordson tractor replacement. United Tractor contracted Allis-Chalmers to build this machine. Due to the United Tractor Company's dissolution by 1930, the machine would later be introduced as the Allis-Chalmers Model U.

22 · The Stylish Engineer

1935 Model 70 · Origin: **Charles City, Iowa** · Company: **Oliver Farm Equipment Company**

▼ The Model 70 was available as a row crop, standard tread, orchard, and industrial machine. FLOYD COUNTY HISTORICAL SOCIETY

The founders of Hart-Parr were two engineering students, and the brand lived on as an engineering-focused company throughout its history. During the dark days of the Great Depression, the Oliver Hart-Parr engineering leader, Oscar Eggen, and a small talented team were hard at work designing a cutting-edge machine. The fruits of their labor, the Model 70, debuted in 1935 as the economy began to rebound. The star of the show was a high-compression engine that was the first six-cylinder in a production tractor, the engineering team also graced the Model 70 with sleek styling and a lovely dark green color .

23 · New Deal Tractor

1938 Caterpillar Farm Crawler D2 · Origin: **Peoria, Illinois** · Company: **Caterpillar Tractor Company**

When Franklin Delano Roosevelt took office in March 1933, he quickly went to work providing relief to the economy with his New Deal. Improved loan terms offered through a variety of government agencies, as well as relief grants, provided access to capital, which in turn helped farmers improve their productivity with a new tractor.

The Caterpillar Tractor Company was created by merger in 1925 and moved the headquarters from California to Peoria, Illinois, in 1930. The smaller machines built by Caterpillar were used in farm work, and the company created the little D2 in response to the New Deal programs. The D2 was first built in 1938, and production ran until 1957.

▲ **A Caterpillar crawler near Meloland, California, digging carrots, which are being picked by migratory workers.** Dorothea Lange / Library of Congress LC-USF34-021056-E

24 · The Country Gentleman

1938 Minneapolis-Moline UDLX · Origin: **Minneapolis, Minnesota** · Company: **Minneapolis-Moline**

Designed to blend the power of a three-plow tractor with the comfort, convenience, and speed of an automobile, the Minneapolis-Moline UDLX was a great idea that was poorly executed and overly expensive to boot. Produced only in 1938, the machine's rarity and gorgeous lines make them highly collectible today.

▲ The Model UDLX was produced in 1938 only, with roughly one hundred sold. The machine had a five-speed transmission that gave it a top speed of 40 mph. Mecum Auctions / Lee Klancher

25 · Swooping into the Orchard

1938 John Deere Model AOS · Origin: **Moline, Illinois** · Company: **John Deere**

The AOS comes as close to sexy as is possible for a vehicle designed to haul carts and sprayers around your orchard or vineyard. That streamlined sheet metal on the flanks has the inglorious job of protecting orchard branches from the tractor's wheels. Created under the supervision of John Deere's engineering leader, Elmer McCormick, the AOS had the gaudiest lines of any Deere tractor in the first half of the twentieth century.

▼ **The Model AO and the AOS are narrow for work in orchards.**
KELLER COLLECTION / LEE KLANCHER

26 · Function over Form

1934 John Deere Model B & 1941 Model HNH · Origin: **Moline, Illinois** · Company: **Deere & Co.**

▶ **A preliminary sketch of the Model B, drafted by Henry Dreyfuss.** COOPER HEWITT, SMITHSONIAN DESIGN MUSEUM / ART RESOURCE, NY

▼ **The Model B was introduced as a more compact, one-row sibling to the two-row Model A. The hood's leaping deer logo was used on only a very few early Model Bs. This example is the first production model, serial number B1000.** KELLER COLLECTION / LEE KLANCHER

In the 1930s, the world was enamored with industrial design. Styled products such as George Grant Blaisdell's timeless Zippo lighter, Walter Dorwin Teague's Brownie camera, and Henry Dreyfuss's Bell telephone became timeless pieces of American culture. The industrial designers became household names, and the products they styled flew off the shelves and boosted company profits.

John Deere saw an opportunity with the Model A and B. Introduced in 1934 and 1935, they were sized perfectly for the small to medium farmers adopting tractors in ever-increasing numbers. They were great tractors, but the look was decidedly analog, a slab of sheetmetal away from the exposed frame rails and componentry of the Waterloo Boy.

In fall 1937, engineering leader Elmer McCormick traveled from Moline to New York City to speak with famed industrial designer Henry Dreyfuss about styling the John Deere tractors. According to the lore, Dreyfuss accepted the job on the spot.

A staunch believer in function over form, Dreyfuss was also an ethical and careful man. He proved a good fit for Deere & Co., and the styled Model A and B were the start of a line of timelessly beautiful machines he designed. The measured approach of Dreyfuss gave John Deere tractors a look that embodied the dedication and care the company invested into its machinery.

▼ **The styling by Henry Dreyfuss graces this Model HNH. Only thirty-seven Model HNH tractors were produced, all between March 11, 1941, and January 23, 1942. This Model HNH, serial number 41760, was built on December 11, 1941, and shipped to Los Angeles.**
KELLER COLLECTION / LEE KLANCHER

27 · Planned Obsolescence

1938 Allis-Chalmers WC Styled · Origin: **West Allis, Wisconsin** · Company: **Allis-Chalmers**

Perhaps the most flamboyant of the 1930s industrial designers, Clifford Brooks Stevens created designs for many major manufacturers, including Jeep and Harley-Davidson. The wide-mouth peanut butter jar and Oscar Mayer's iconic Wienermobile can be credited to his genius as well.

Stevens pioneered the concept of "planned obsolescence," meaning new designs should be used to make new models more desirable, and to make the old ones appear to be obsolete. That concept is so ingrained in modern product marketing that it seems obvious today.

Based in Wisconsin with offices not far from West Allis, he was the logical choice to add some style to the Allis-Chalmers line. When Stevens took his first drawings of a styled tractor to the engineering team at Allis-Chalmers, the response was uninspired.

"What man worries about how a tractor looks?" the Allis engineer responded. "If it plows the field, that's enough."

He was wrong and, thankfully for Persian Orange fans, over-ruled. The Brooks Stevens styling of the Model WC created one of the most popular Allis-Chalmers tractors ever built.

▼ **The Allis-Chalmers Model WC was built from 1933 to 1948. The styling you see below debuted on the 1938 model.**
Gary Nelson

▲ The Model WC was styled by famed industrial designer Brooks Stevens. This circa 1960 image shows his office.

◄ **Early Brooks Stevens sketch.**

28 · The Art of the Sale

1939 Farmall M & H · Origin: **Chicago, Illinois** · Company: **International Harvester Company**

▶ **Industrial designer Raymond Loewy was commissioned to sketch the new Farmall tractor. Flamboyant, talented, and famous, Loewy also designed the Studebaker, Air Force One, and many company logos including Lucky Strike and Exxon.**

Gregg Montgomery Collection

Hard times test the mettle of good companies, and the Great Depression was a challenge that IHC handled with aplomb. Despite minimal sales and grave losses on the balance sheets in the early 1930s, the company had so much cash on hand that it continued to pay full dividends to stockholders. When IHC laid off employees at that time, it offered many of them loans to tide them over until they could return to work.

The company also responded to the farm crisis by putting its formidable engineering teams to work developing an all-new line of tractors.

When the new line was first introduced in 1939, the world's economy had only modestly recovered, and life on the farm was still hard. The Farmall M and H set this still-tough market on fire. They were smartly engineered, brilliantly marketed, and sold through the most powerful agricultural equipment sales network in the industry.

The crowning touch was the timeless lines drawn by the famed industrial design firm of Raymond Loewy, and the straight and true machines are functional and attractive even today.

Sales were strong in 1940 and 1941, slowed down the next two years, and then steadily increased until peaking in 1948, a time when tractors sold in unparalleled numbers. According to US Census data, tractor production hit an all-time high of 753,623 units built in 1948.

During that year, IHC grew its market share to 23.4 percent—nearly 10 percent more than the nearest competitor, John Deere. The H and M carried a large amount of that load, combining to sell nearly 800,000 units in their production span from 1939 to 1954.

Although the original Farmall was the most innovative red tractor in the first half of the twentieth century, the Farmall H and M represented the pinnacle achievement of IHC's engineering prowess, marketing savvy, and unsurpassed sales network.

◀ The Farmall H and M designs were a startling departure from the mechanically and visually disjointed predecessors. This sketch shows how the basic design evolved. GREGG MONTGOMERY COLLECTION

▲ The Farmall H was the most prolific tractor in history, and total production is estimated at just more than 420,000 sold (this is both H and Super H production). JIM JOHNSON COLLECTION / LEE KLANCHER

29 · Henry's Return

1939 Ford Model 8N · Origin: **Detroit, Michigan** · Company: **Ford Motor Company**

▼ **Like the Fordson, the Ford 8N disrupted the tractor industry, and it was produced from 1939 to 1942. This Ford is at work on the Manzanar Japanese Relocation Center in California in 1943.** Ansel Adams / Library of Congress LC-DIG-ppprs-00119

While Henry Ford essentially left the American tractor market in 1928, the Fordson lived on and was produced in Cork, Ireland. The models sold well internationally, and were imported to America in very limited numbers.

In the 1930s, Ford looked to shake up the American market again. As with the Fordson, his strategy was to offer a tractor that was light, high-tech, and cheap. His $585 Model 8N checked all those boxes and was sharply styled as well. The color was a tad dull, but as we know all too well, Ford preferred to keep colors simple.

That said, Ford had an ace up his sleeve. One of the most significant innovations of the early twentieth century was the three-point hitch that debuted on the 8N. The hitch used the Ferguson System, which used leverage to press down on the front wheels when the tractor had a heavy load pulling on the drawbar. This

July 4, 1933.

H. FERGUSON

1,916,945

TRACTOR DRAWN AGRICULTURAL IMPLEMENT

Filed June 6, 1929

2 Sheets—Sheet 1

Inventor:

H. Ferguson

Atty.

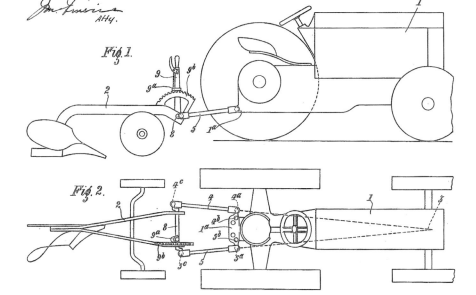

Fig. 1.

Fig. 2.

◀ **The Ferguson System on the N Series tractors was developed by Harry Ferguson.** US PATENT

▼ **This Ford tractor was purchased by the Hansen-Frederickson Machinery Cooperative of Sanpete County, Utah, with an FSA loan. Mariel O. Hansen is driving the tractor.** LIBRARY OF CONGRESS LC-USF34-014106-D

THE RISE OF STYLE   73

▲ Industrial designer John H. Walters was hired by Edsel Ford in 1935, where he worked for influential designer Eugene "Bob" Gregorie. Walters made this rendition of a Ford tractor on April 13, 1939. The idea was to create a new appearance for the machine, driving demand without high research and development costs. Walters worked for Ford Until 1941, when he started his own design firm.

Cooper Hewitt, Smithsonian Design Museum / Art Resource, NY

eliminated the deadly tendency of the Fordson (and other early tractors) to flip over when the plow or other implement hit a rock or other obstacle.

The innovative hitch system was developed by British engineer Harry Ferguson, who wanted Henry Ford to see the machine. He flew himself and his design to Ford's Michigan estate, Fair Lane. After five days of demonstrations, Ferguson and Ford sat down at a table and made a handshake agreement that Ford would create a new tractor equipped with Ferguson's system.

The first model to use the system was the Ford 9N, which was produced from 1939 to 1941. The model evolved into the 2N, produced from 1942 to 1947, and the 8N, produced from 1947 to 1952. These models are referred to as the "N-Series" Ford tractors.

Ferguson and Ford acrimoniously split in 1947. The three-point hitch remains the dominant form of tractor hitch used today.

▲ The Ford Model 8N was the last generation of the N Series tractors, and was produced from 1947 to 1952. The distinctive red engine and transmission made it one of the best-looking tractors. Lee Klancher

30 · Fighting for Position

1954 Case Model SC · Origin: **Racine, Wisconsin** · Company: **J. I. Case Company**

In the 1920s, J. I. Case had all it could do to maintain the number three position in the industry. The company appointed Leon Clausen, who had been a high-level manager in the railroad industry and for John Deere. Opinionated, driven, and a micro-manager, Clausen was the burning furnace at the heart of the company. He drove change and helped them compete—he also enforced his opinions as the rule of law. For example, the company improved sales by ditching the drab gray paint in favor of a vivacious Flambeau Red in 1938, but Clausen later ignored complaints that the color was too close to Persian Orange.

One of the hot trends of the late 1930s was small one-plow tractors, and the Allis-Chalmers B and Farmall A and B were the market leaders. Clausen was reluctant to enter the small tractor market due to the thin profit margins. He also believed the Ferguson draft control system was foolish. His opinions slowed progress, although they did eventually get into this market with the Model V series tractors.

The Model S was a response to the market designed to compete head-to-head with the Farmall H, and it did so with aplomb. The machine combined an improved, higher-rpm engine with the layout and design from the Model D. The Model SC was the row-crop version of the Model S.

Clausen's interest in controlling every aspect of engineering was such that he left behind a half-day's correspondence with an engineer about whether or not to include a fuel sediment bowl that cost pennies. He also liked to argue with salespeople, telling them that the equipment was exactly what they needed and their job was to sell.

The legendary chicken roost steering—the steering control rod that jutted out awkwardly on the early Case tractors—was said to have been held in production because Clausen did not have an issue with it.

Under his leadership, however, J. I. Case gained more than it lost. By 1949, the company had introduced a proprietary draft control hitch (the Eagle Hitch), and was sharpening its sales pitch to offer the "Big Bargain" in tractors. Offering smartly engineered machines at prices lower than the competition worked, as J. I. Case increased its market share from 3.7 percent in 1944 to 7.0 percent in 1948.

▼ **The Ferguson three-point hitch was patented, and competitors had to design a similar system. Case had the "Eagle hitch."** Case IH

▶ **The Case Model SC was the row-crop version of the Model S and was produced from 1941 to 1954. This is a 1954 example, and it has lived on the same square mile its entire life.** Wakeman Collection / Lee Klancher

WAR PRODUCTION

Tractor manufacturers played important roles during World War II. The machines were in demand to help the depleted labor pool up their productivity and feed the world. Materials to build agricultural equipment were hard to come by, however, and often only obtained through government programs.

Tractor manufacturers scrounged and begged for materials and built as many tractors as they could, while the bulk of their manufacturing operations built tanks, guns, bombs, and parts for the war effort. The world's heavy equipment manufacturers—tractor makers among them—proved to be a vital asset during the war.

▲ Tractors were used by the military, particularly industrial models like this International I-4 that is pulling a B-25 Mitchell bomber at North American Aviation, Inc., in Kansas City in October 1942.

Library of Congress LC-DIG-fsac-1a35288

◀ Women around the world ran farms and operated the equipment during World War II. These women harvesting beets are members of the British Women's Land Army. LIBRARY OF CONGRESS, BRITISH MINISTRY OF INFORMATION, LC-USW34-000622-ZB

▼ The tractor was promoted as a tool to help the war effort in this 1942 poster. FRED CHANCE, OFFICE OF WAR INFORMATION, NATIONAL ARCHIVES

THE POST-WAR BOOM
1945–1953

The most prolific era in tractor production came just after World War II. Tractor technology was mature, farms were rapidly adding more acreage, and the farm economy grew explosively.

31 · The Last Solo

Cletrac Crawler · Origin: **Cleveland, Ohio** · Company: **Cleveland Tractor Company**

Agricultural companies rarely survived with just one equipment type, and most floundered during the Great Depression. Cletrac, with roots that dated back to 1899, ran counter to both trends. Armed with progressive engineering and sharp management, Cletrac thrived by building smartly designed crawlers in the viciously competitive tractor market of the 1920s as well as the disastrous market of the 1930s.

During World War II, Cletrac not only supplied the US military with crawlers, it also subcontracted work to John Deere to build Cletrac MG-1 military tractors in March 1942. According to author Wayne G. Broehl Jr., Deere leader Charles Wiman saw this as a great opportunity and proposed to the board

▲ The Cletrac 30B was introduced in 1929 at the dawn of the Great Depression. Production continued until 1930, with approximately 1,450 built in total. Super T © 2017

that Deere merge with Cletrac. Shortly after making this suggestion, Wiman was asked to serve in the Ordnance Corps of the US military. He answered the call to duty and played a significant role, and would later help run the farm machinery and equipment arm of the War Production Board with IHC executive Harold Boyle. The Cletrac merger was sidelined, as Wiman was the primary driver behind it.

In October 1944, Cletrac was purchased by the Oliver Farm Equipment Company, who added the crawlers to its line and also changed the company name to the Oliver Corporation. The crawlers remained in production until 1965.

32 · The Consolation Prize

John Deere Model BO Lindeman Crawler · Origin: **Yakima, Washington** · Company: **Lindeman Manufacturing, Inc.**

▼ The John Deere Model BO Lindeman Crawler was built from 1936 to 1947, with about 1,700 produced. Lee Klancher

The Lindeman brothers—Harry, Jesse, and Ross—of Yakima, Washington, owned a variety of dealerships before settling on John Deere in the 1930s. They soon discovered that their area needed a crawler that Deere didn't make, so they built tracks for a John Deere GP tractor and shipped it to the headquarters in Moline for testing. John Deere management approved, and the brothers found the newer Model B to be a better donor for their creation. They built and sold these with good success for more than a decade.

In 1947, with Deere & Co. flush with cash and president Charles Wiman eager to expand, the company purchased Lindeman Manufacturing, Inc.

Jesse Lindeman stayed on to develop a new crawler, the John Deere Model MC.

33 · The Beast

1947 International TD-24 Crawler · Origin: **Chicago, Illinois** · Company: **International Harvester Company**

In 1947, the International Harvester Company remained the largest agricultural company in the world, and it acted the part. Under the leadership of Fowler McCormick, the company looked to expand into a variety of new markets (while its tractor line began to get long in the tooth).

One of the midcentury IHC visions was to compete with Caterpillar in the construction industry. IHC's grandest entry into that market was the TD-24, which was the world's largest and most powerful bulldozer. Rushed into production, the machine had near-fatal issues with the driveline. By the time those were fixed, the reputation of the machine was badly sullied and a mind-boggling investment in research and development was wasted.

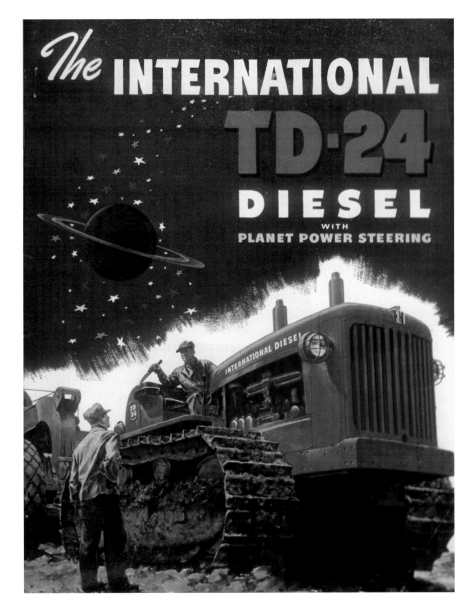

◀ **The International TD-24 Crawler was produced from 1947 to 1959.** Wisconsin Historical Society 10086

34 · The Beauties

1948 Oliver Models 66, 77, and 88 · Origin: **Charles City, Iowa** · Company: **Oliver Corporation**

▲ **The new-for-1947 line of Oliver Fleetline tractors included the models 66, 77, and 88.** Lee Klancher

In 1942, **Oliver** was working on an upgrade to the successful line of tractors, and turned to industrial designer Wilbur Henry Adams for help. Adams was trained at MIT, and after working for Raymond Hood in New York and assisting with the design of the Rockefeller Center, he returned to his native home of Erie, Pennsylvania. He built a studio in an old barn, where he penned his designs while tending to chickens and other animals on his farm.

An Oliver 60 was delivered to Adams's farm, which he then used as a live model, heaping it with clay to restyle the body and grille. He also sent sketches to Oliver engineer Herbert Morrell, who worked with Adams to translate the designs into manufacturing plans. "Wilbur's designs were beautiful and exotic for the time," Morrell wrote in his book, *Oliver Tractors*. "I learned very much from him."

The new line also featured new engines, which Oliver had built by Waukesha, and new transmissions, seats, and much more. The line included an optional diesel engine, which was one of the best on the market at that time.

One of the interesting sidelines of the machine was the independent power take-off (IPTO), which allowed the power take-off to continue to run even when the tractor's clutch was depressed. This was a critical development, and a rarity in the late 1940s. According to Morrell, Oliver released a pilot run of 300 Model 88s that had an IPTO in 1946.

As an aside, some have claimed that the Cockshutt Model 30 had the first IPTO. That's incorrect. According to Morrell, the first IPTO appeared on the 1928 Hart-Parr 18-36.

▶ **The Oliver 88 was introduced in 1947 with this styling. In 1948, the model (along with the 66 and 77) was treated to new paint and styling.** Lulich Collection / Lee Klancher

▶ This sketch was done by industrial designer Wilbur Henry Adams, who was contracted by Oliver to design the Fleetline tractors. Adams was a tremendous talent, but never designed a cultural icon that made him famous.

35 · The Stylish Alternative

1952 Massey-Harris Pony 11 · Origin: **Toronto, Ontario** · Company: **Massey-Harris Company Limited**

▶ **The Massey-Harris Pony 11 was built from 1947 to 1957.**

Colburn Collection / Lee Klancher

Small tractors well suited for vegetable farmers or industrial work were popular in the late 1940s, and the Massey-Harris addition was the Pony. The small machine had a high-compression, 62-cubic-inch four-cylinder engine rated for 11.62 horsepower.

36 · A Small Sensation

1947 Farmall Cub · Origin: **Chicago, Illinois** · Company: **International Harvester Company**

▶ **The 60-cubic-inch four-cylinder put out 9 horsepower and featured a dizzying array of attachments. This version is an International Cub Lo-Boy, produced in the 1950s.**

Sorenson Collection / Lee Klancher

In April 1947, production began of the all-new Farmall Cub, produced through 1964. The machine proved to be a sensation, with more than 65,000 built in the first twenty-three months of production.

A *Life* magazine feature depicted Louisville Works building 2,200 small tractors a week during this time.

37 · A Tidy Package

1947 John Deere Model M · Origin: **Moline, Illinois** · Company: **Deere & Co.**

John Deere weathered the fiscal storms of the 1940s with its usual caution, and introduced only mildly updated models from 1939 to 1946. The new-for-1947 M was the company's first new model in eight years. An all-new, one-row tractor with 20 horsepower, hydraulics, PTO, electric starting, and clean Dreyfuss styling, the M was a strong entry into the small tractor market. The first one was delivered to the ranch of company leader Charles Deere Wiman on April 1, 1947. In 1949, the Model M became the first tractor produced at John Deere's new Dubuque Tractor Works.

◀ **The John Deere M was built from 1947 to 1952. More than eighty-eight thousand were produced.** Vinopal Collection / Lee Klancher

38 · The Innovator

1948 Allis-Chalmers Model G · Origin: **West Allis, Wisconsin** · Company: **Allis-Chalmers Manufacturing Company**

Although the rest of the tractors on this page were built to do it all, the Model G focused on cultivating and planting. With a wide array of attachments, the machine quickly became popular with crop and vegetable farmers. The 62-cubic-inch Continental engine made 10 horsepower, and was the same engine used in the Massey-Harris Pony. The tractor remains popular today with small farms, and a good number of Model Gs have been converted to use electric engines.

◀ **The Model G was produced at the A-C plant in Gadsden, Alabama, from 1948 to 1955, with more than twenty-nine thousand built.** Mecum Auctions / Lee Klancher

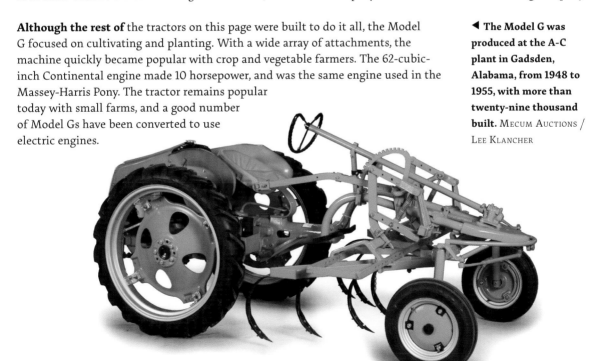

THE MARSHALL PLAN TRACTORS

The Marshall Plan was an American initiative to provide relief to Western Europe after World War II. Beginning in April 1948, it provided $13 billion in economic assistance. The plan helped rebuild war-torn areas by reducing trade barriers and encouraging modern business practices. The UK, France, and Germany were the largest recipients of aid.

The French placed a high emphasis on rebuilding their agriculture, and a shipment of tractors was purchased in 1948 by the French using Marshall Plan funds.

▲ These Massey-Harris 44 tractors were delivered to France as part of the Marshall Plan, and the photographs are dated January 1949. National Archives 286-MP-fra-00274a

▲ The Massey-Harris tractors were delivered with a hand-lettered sign, reading, "We wish you a Merry
Christmas and a Happy New Year from the people of the USA to the people of France!" NATIONAL ARCHIVES 19973973

39 · The Frugal Twin

1953 John Deere Model R · Origin: **Moline, Illinois** · Company: **Deere & Co.**

A fuel-efficient diesel was a goal for the tractor industry in the 1940s, and John Deere took its time getting this issue absolutely right. The Model R had one of the longest development cycles of any John Deere tractor, and getting the large displacement twin running well was a massive challenge. When it was released to the public in 1949, the new model had spent fourteen long years in development and more than sixty-six thousand hours being field-tested. The engineering team coupled a high-compression engine with a small V-4 gasoline-powered starting motor. Start the small motor, and it cranked the big diesel up to speed, cold, hot, or otherwise. The Model R set an all-time record for fuel economy when tested at the University of Nebraska in 1949.

◀ **The Model R was introduced in 1949 and produced until 1954. This original condition machine was sold at the Keller family dealership in 1953.**
KELLER COLLECTION / LEE KLANCHER

40 · The Shotgun Prophecy

1949 Field Marshall Series 2 · Origin: **Gainsborough, Lincolnshire** · Company: **Marshall, Sons & Co.**

Starting big diesel engines was problematic in the mid-twentieth century. The Field Marshall engineers had a novel solution for starting their machine's single-cylinder diesel: they used a shotgun shell.

A casing with no shot was placed in a receptacle, the engine was rotated to top dead center, a decompression valve was opened, and the operator hit the receptacle (and the shell) with a hammer. With a bang, the engine turned over and started.

Rumor has it the company staked their future on this innovation. They went out of business in the early 1960s.

◀ **Field Marshall Series 2 tractors were built from 1947 to 1949.**
MECUM AUCTIONS

41 · A Streamlined Vision

Minneapolis-Moline ZAS · Origin: **Hopkins, Minnesota** · Company: **Minneapolis-Moline**

One of the sleekest Minneapolis-Moline tractors was the Model Z. The narrow front ZA and wide front ZAS were upgraded versions of the Z introduced in 1949. The company dubbed itself "modern tractor pioneers," and touted releasing the first modern cab as well as the first "visionline" tractors. The progressive company was one of the first to use four-color advertising.

▶ **The Minneapolis-Moline ZAS was built from 1949 to 1953.** Lee Klancher

42 · The Big Boy

Lehr Big Boy · Origin: **Shelbyville, Illinois** · Company: **Custom Manufacturing Company**

The Custom Manufacturing Company of Shelbyville, Illinois, used Chrysler six-cylinder engines in their tractors, and sold them under a number of different brands, including Lehr and Custom. The Lehr-branded machines were sold by Lehr Equipment in Richmond, Indiana.

◀ **The Lehr Big Boy is a rebranded Custom Model C, which the owner estimates was built in 1948.** Haecherl Collection / Lee Klancher

THE UNI-TRACTOR

1951 Minneapolis-Moline Uni-Tractor · Origin: **Hopkins, Minnesota** · Company: **Minneapolis-Moline**

By Larry Gay

New for 1951, the Minneapolis-Moline Uni-Tractor was a unique tractor designed to carry harvesting machines instead of pulling them. The tractor was built with the two drive wheels in front. This configuration left the space above and behind the front axle open for a combine or corn picker attachment, making the Uni-Tractor the propelling unit of an interchangeable, self-propelled harvesting machine named the Uni-Harvester.

The machine became a self-propelled, two-row corn picker when the Uni-Husker was attached. The corn picker featured a husking bed with ten 36-inch-long husking rolls. The combine and corn picker attachments were derived from the company's

69 combine and two-row, pull-type corn picker. When the harvesting attachments were exchanged, a hoist was provided to lift the unit, and a three-wheel dolly was used to hold it for storage. A picker-sheller, a hay baler, a forage harvester, and a windrower for the self-propelled Uni-Harvester system were added later.

Several different models of Uni-Tractor were produced by Minneapolis-Moline until 1963, when New Idea purchased the rights. The machinery was then renamed the Uni-System, and New Idea produced a wide variety of models ranging from 38 to 234 horsepower. The Uni-System was discontinued in 1993, when AGCO acquired White–New Idea.

▲ The Uni-Tractor was powered by a 206-cubic-inch Minneapolis-Moline V-4 gasoline engine with a rating of 38 engine horsepower. The transmission was a three-speed gear box with one speed in reverse. The belt drive for the harvesting units was connected to the right end of the engine crankshaft and was independent of the traction drive. The loader in the picture is a concept that was not produced. MINNESOTA HISTORICAL SOCIETY

▲ This late model Uni-Tractor is equipped with a forage harvester. A Uni-Combine option was also offered. Minnesota Historical Society

▲ The hay baler was added later to the Uni-Tractor. Ground speeds ranged from 0.9 mph in first gear to 9.8 mph in third gear. Minnesota Historical Society

43 · Supers

1953-54 Farmall Supers · Origin: **Chicago, Illinois** · Company: **International Harvester Company**

By 1953, International's large Farmall H and M had been on the market for fourteen years. Farms were getting bigger, and operators needed more precise speed control, improved implement control, and higher horsepower. Management's focus was on growing the company with a broader product line rather than maintaining its stranglehold on the tractor industry. The industry, in the meantime, was innovating.

The Farmalls desperately needed a major upgrade, and the Super models that appeared in 1953 added more power and a few key upgrades. In 1954 only, the Super MTA introduced the torque amplifier, a two-speed gear-range transmission added to the regular one that could be shifted on the fly.

While the Supers were immensely popular tractors, the fact that this top-selling line was left fallow for nearly a decade and a half left the door open for other manufacturers to catch up.

▼ **This is the Super M, which offered a host of new features and was built from 1952 to 1954.** Olson Collection / Lee Klancher

◄ The Farmall Super H was produced in 1953 and 1954. This one is pictured at work with a No. 64 combine in 1953.

Wisconsin Historical Society 8864

◄ This is the Super BMD, a British diesel version of the model, which was built in Doncaster, England, from 1953 to 1959.

Lee Klancher

◄ The final variant of the M was the Super M-TA, which was built only in 1954 and was equipped with a torque amplifier, a two-speed range transmission that could be shifted on the fly. This is a diesel version, the Super MD-TA.

Lee Klancher

HUNGRY FOR POWER
1954–1959

By the mid-1950s, the farmer's need to work large plots of ground was driving demand for high-horsepower tractors. The major manufacturers were behind, but innovative independents would step up and lead the way. CASE IH

44 · The Spark

Wagner TR14 · Origin: **Portland, Oregon** · Company: **Wagner Tractor, Inc.**

Perhaps the most important tractor of the 1950s was the Wagner, a four-wheel-drive machine that satisfied power-hungry farmers and whose seemingly overnight success would serve as an industry-wide alarm bell. While the major builders focused on boosting power from 60 to 70 horsepower, the Wagner brothers unveiled three new articulated four-wheel-drive tractors with 114 to 165 horsepower in 1954. By the early 1960s, Wagner Tractors would be a hot item with farmers working large acreage, an inspiration to independent-minded builders, and a source of aggravation for management teams at established tractor manufacturers.

▶ **The Wagner was one of the first popular high-horsepower four-wheel-drive tractors.**

PETER SIMPSON
COLLECTION

▶ **Elmer Wagner filed this patent for his four-wheel-drive tractor in September 1953.**

US PATENT

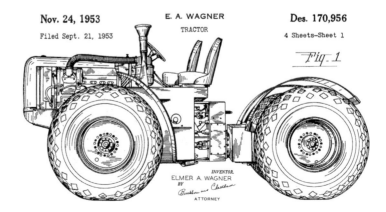

Nov. 24, 1953   E. A. WAGNER   Des. 170,956

Filed Sept. 21, 1953   TRACTOR   4 Sheets–Sheet 1

Fig. 1

INVENTOR.
ELMER A. WAGNER
BY
ATTORNEY

45 · The Truck Builders

FWD Experimental Tractor · Origin: **Clintonville, Wisconsin** · Company: **Four Wheel Drive Auto Company**

The first successful four-wheel-drive car was built in 1908 by the Four Wheel Drive Auto Company, better known as FWD. The company's innovative four-wheel-drive machines were built for military as well as civilian use, and expanded to have subsidiaries in Canada and Britain. Although its primary market was trucks, it experimented with tractors. These experiments became a market expansion in 1961—see tractor number 62—and the FWD name lived on into modern times as a builder of firefighting machines. FWD is still based in Clintonville, Wisconsin.

▼ **The four-wheel-drive tractor began to emerge in the 1950s. This early experimental four-wheel-drive tractor was built by FWD.**
Wisconsin Historical Society 129279

46 · Compression Wizardry

1954 Oliver XO-121 · Origin: **Charles City, Iowa** · Company: **Oliver Corporation**

▼ **The Oliver XO-121 was an experimental machine built to test high-compression diesel engine performance.**
Floyd County Historical Society

Oliver was one of the most progressive tractor companies of the 1950s, and one of their interesting developments was the high-compression engine built in the XO-121 experimental tractor. New fuels were available, and Oliver management was interested in using them in high-efficiency, high-compression diesel engines. Oliver cooperated with Ethyl Corporation, a fuel additive company, to create a diesel engine that used a compression ratio of 12:1 to improve performance and fuel economy. The machine that housed the engine was the XO-121, and the performance proved remarkable, thanks to some combustion chamber wizardry.

The film, *Getting Ahead of Tomorrow,* told the story of the engine's creation. Project engineer Herbert T. Morrell presented his findings at a meeting of the American Society of Agricultural Engineers. Morrell recalled speaking to his friend Stanley Madill, a top engineer for John Deere, who told Morrell, "You will never know what you just have done for John Deere Engineering." The new-for-1961 Next Generation John Deere tractors featured high-compression engines.

The blackboard reads:

EFFICIENT POWER   HIGH COMPRESSION
FLEXIBLE POWER   HIGH OCTANE FUEL
ECONOMY

LONG LIFE   RUGGED DESIGN

▲ The team of Oliver management and engineers planning the **XO-121.** Floyd County Historical Society

◄ **Oliver chief engineer Herb Morrell (standing) at work.** Floyd County Historical Society

THE STAR AND THE FARM

Actor and cultural icon James Dean spent a good portion of his formative years on the farm of his uncle, Marcus Winslow, in Fairmont, Indiana. He moved off the farm to pursue his acting career, but would return on occasion. These images were taken in 1955 by photographer Dennis Stock, who met Dean just as *East of Eden*, the first film to feature Dean, was released. Stock recognized Dean's talent, and immediately began photographing him. When Dean told Stock about growing up on a farm in Indiana, Stock convinced *Life* magazine to send him out with Dean to photograph his roots on the farm.

Life agreed to pay Stock $300 for two days of making photographs. Stock, though, dedicated two months shooting images of Dean in New York, Los Angeles, and Indiana, and the spread in *Life* that resulted helped cement Dean's position as a rising young American actor.

Dean made three popular films and was killed in a California car accident in September 1955. The images that Stock took on the farm have become a part of the legend surrounding Dean.

The farm and the tractor in the image remain much as they were. Coy Winslow, the descendant of Marcus and one of the family caretakers of the farm, is also a tractor collector.

In February 2016, the *New York Times* published a video interview of Stock about his time with Dean, and Stock said the images on the farm were a key part of the spread that ran in *Life* magazine. He also said his favorite times working with Dean took place in Indiana. "The aunt and uncle were wonderful people," Stock said. "The best memories I have of him are at Fairmont."

▶ **The farm provided photographer Dennis Stock an opportunity to capture Dean in candid moments.**
Dennis Stock / Magnum Photos

▶ ▶ **James Dean grew up on this farm in Indiana, and this Minneapolis-Moline tractor remains on that farm in good condition.**
Dennis Stock / Magnum Photos

47 · A Welder and a Dream

1957 Steiger No. 1 · Origin: **Fargo, North Dakota** · Company: **Steiger Manufacturing, Inc.**

▼ Douglass, Maurice, and John Steiger built this No. 1 by welding, bending, and cutting the frame, bodywork, and more, to fit their design around a used engine and transmission. The tradition of high-quality metal work carried with the company for decades, and a global brand was born. BONANZAVILLE MUSEUM COLLECTION / LEE KLANCHER

The first Steiger tractor was inspired by a Wagner, authorized by an optimistic banker, and built by a hands-on family of farmers.

John Steiger and his sons, Douglass and Maurice, grew up farming in far northwestern Minnesota. Struggling with low farm prices and heavy rain, the Steigers survived by running heavy equipment for cash. The farm required two or three men to run the family's 40-horsepower International tractors during harvest. The Steigers became aware of the 100+ horsepower Wagner tractor, and saw that with such a machine, John Steiger could do most of the field work on his own, and the boys could focus on contract work to keep the cash flowing.

Douglass asked the president of the Union State Bank in Thief River Falls for a loan of $20,000 to buy a new Wagner. He also suggested that with a loan of only $10,000, he and his family believed they could make such a machine in their shop.

The banker told them to build rather than buy.

In the fall of 1957 and winter of 1958, the Steigers worked day and night in their barn they had converted into a shop to create Steiger No. 1, wrapping a hand-built frame, power divider, and more around a 426-cubic-inch GM six-cylinder diesel pumping out 200 horsepower.

▶ **Douglass Steiger and long-time employee Clifton Johnson in the early 1960s.**

None of them had formal training in engineering nor more education than a high school diploma, but they had built implements and were extremely gifted when it came to metalwork.

The machine that resulted was Steiger No. 1, and it worked effectively on the family farm for well over a decade. The machine is alive today, and the layout of the controls is logical and well thought out. The family started building smaller machines for the neighbors, and in two decades, their green machines would grow to become an industry-leading, high-technology four-wheel-drive tractor.

▲ **Douglass Steiger and No. 1 at the Big Iron Show in Fargo, North Dakota, on September 13, 2017.** OCTANE PRESS ARCHIVES

48 · The Art of Refinement

1958 John Deere 830 · Origin: **Moline, Illinois** · Company: **Deere & Co.**

The handsome 830 was the penultimate iteration of the brilliantly engineered Model R. Technological refinement and a splash of color kept the Model 830 more or less current with the market. The engine remained extremely fuel efficient, and with 75 belt horsepower, was one of the more powerful offerings from the major makers in 1958.

The big drawback of the machine was the V-4 starting engine, which added complexity and the need for two fuel tanks to fill.

Even so, the 830 was one of the best high-horsepower tractors on the market at the time. The problem was that while the established companies rolled out new color schemes, body work, and refinement, people like the Steigers and Wagners were stuffing 150-plus horsepower engines into home-built frames to build the machines that large acreage farmers wanted and needed.

▼ **This Model 830 is the first production model, serial number 830000, and was constructed on August 4, 1958.**
KELLER COLLECTION / LEE KLANCHER

49 · Red Flags

1962 International 560 · Origin: **Chicago, Illinois** · Company: **International Harvester Company**

▲ **1962 Farmall 560 and International 82 combine.**

ARMSTRONG COLLECTION / LEE KLANCHER

By 1958, nearly two decades had passed since the International Harvester Company (IHC) had launched an all-new line of tractors. They were overdue for exciting and new, and pulled out all (or most of) the stops for the new-for-1958 40 and 60 series machines. The sheet metal design and paint scheme were sharp and modern, and the feature set carefully tailored to the extensive surveys of IHC customers.

The machines were burdened with several problems. One issue was that the company rushed production and the rear end on the 560 wasn't quite up to handling the engine's horsepower. The result was the largest recall in company history, with thousands of tractors being fitted with upgraded rear ends.

The rear end problem was an expensive, embarrassing mistake, but it wasn't the big problem. IHC sold the machine briskly and it was, overall, a well-engineered and effective machine with an optional up-to-date, electric start diesel engine. The more serious issue was that the tractors weren't a technological leap, particularly in the ease of operation and horsepower output.

The competition was hard at work building a machine bristling with technology and style, a machine that was being tested as the 560 was launched.

Too little, too late is the story when it comes to the 560.

50 · Chasing the Market

1959 John Deere 8010 · Origin: **Moline, Illinois** · Company: **Deere & Co.**

▼ This Model 8010, serial number OW41A, was positively identified by John Deere 8010 engineers Gerry Mortensen and Deryl Miller as the first prototype used by the John Deere Engineering Research Division in Cedar Falls, Iowa. The men confirmed this tractor also appeared at John Deere Day in Dallas, with a 1010 tractor on the rear hitch, but did not appear at the introduction in Marshalltown, Iowa.

<small>Keller Collection / Lee Klancher</small>

At a **1959 John Deere** field day in Marshalltown, Iowa, a brand-new machine shocked the crowd: a 200-horsepower, four-wheel-drive behemoth called the John Deere 8010.

The powerful machine had features uncharacteristic for Deere.

For starters, the engine was not a John Deere two-cylinder, but a 425-cubic-inch, six-cylinder, two-stroke Detroit Diesel. The transmission was a Spicer nine-speed, with Clark axles and Westinghouse air brakes. The tractor also featured a shocking $33,000 price tag.

The outsourced bits and sudden appearance made it clear this tractor was a specification build, which meant an engineer drew up the design using as many off-the-shelf parts as possible. This allowed for a new model to be created rapidly, but required compromises based on the available parts and pieces.

John Deere favored long development cycles—think of the fourteen years it took to make the Model D as a classic example. The off-the-shelf parts on the 8010 indicates the machine was created with a short development time. The most likely presumption is the machine was a response to the Wagner tractor, or at the worst the increasing market demand for high-horsepower machines.

The testing time was clearly inadequate. The Spicer transmissions in the 8010 failed under significant use, a flaw that would have appeared in a normal test cycle.

John Deere built one hundred of the machines, and only a handful were sold. In the end, all but one of the one hundred 8010s were returned to Deere in a recall. These ninety-nine machines would reappear several years later as 8020s, with just enough improvements to entice a few customers to take them home.

The one remaining prototype is pictured here: a twenty-thousand-pound survivor of a hastily conceived program to meet the power-hungry needs of midcentury big acreage farmers.

THE SPACE AGE
1960–1969

In the 1960s, an industry-wide race to increase power and introduce new technology prompted tractor manufacturers to build some of the most influential machines of the twentieth century. WISCONSIN HISTORICAL SOCIETY

51 · State-of-the-Art Styling

1960 Minneapolis-Moline M-5 · Origin: **Hopkins, Minnesota** · Company: **Minneapolis-Moline**

In addition to penning tractor designs for Allis-Chalmers, Brooks Stevens also drew for Minneapolis-Moline. This angular restyling of the line sported a look that was well ahead of its time, and the M-5 that resulted gave the fans of Prairie Gold a crisp, fresh look.

▲ **Brooks Stevens's concept drawings for the Minneapolis-Moline M-5. The production tractor was offered from 1960 to 1963.** MILWAUKEE MUSEUM OF ART / BROOK STEVENS COLLECTION

▶ The Minneapolis-Moline M5 was produced from 1960 to 1963, and was available with gas, diesel, and LP fuel engines. The gasoline version produced 61.0 PTO horsepower when tested at Nebraska in 1960. MINNESOTA HISTORICAL SOCIETY

52 · Fuel Cell Future

1959 Allis-Chalmers Fuel Cell Tractor · Origin: **West Allis, Wisconsin** · Company: **Allis-Chalmers Company**

▲ Allis-Chalmers began experimenting with fuel cell power in the early 1950s, resulting in this experimental tractor. The machine plowed a field of alfalfa with a two-bottom plow in October 1959, and produced no heat, smoke, or noise.
SMITHSONIAN AG.76A8

A fuel cell generates electricity by creating a chemical reaction in a solution. The fuel is a solution, which can be pure hydrogen, water laced with potassium chloride, or other gases or liquids. The concept has been around since Sir William Grove cooked up the idea in 1838, and people are still trying to make the technology commercially viable. NASA's Apollo spacecraft was powered with alkali fuel cells (which burn pure hydrogen), as was this Allis-Chalmers experimental tractor. The machine used 1,008 fuel cells to send 15,000 watts to an electric motor, and could pull a weight of 3,000 pounds. The machine was donated to and is owned by the Smithsonian Museum.

53 · The Turbine Terror

1961 HT-340 · Origin: **Chicago, Illinois** · Company: **International Harvester Company**

When the International Harvester Company purchased the Solar Turbines Company in 1959, it soon began experimenting with a Solar-built turbine engine in a tractor, creating this HT-340. The turbine engine never made production, but the hydrostatic transmission did. Several running experimental HT-340s were built and shown around the country, and one of which now resides at the Smithsonian Museum. Ford also built a turbine-powered tractor, the Typhoon.

◀ **The experimental HT-340 was powered by a Titan T62T gas turbine engine that ran at 57,000 rpm, produced 85 horsepower, and weighed only 90 pounds.**
GREGG MONTGOMERY COLLECTION

◀ **The experimental HT-340 being tested. The machine was first shown to the public in 1961 at the University of Nebraska's 10th Annual Tractor Day.**
WISCONSIN HISTORICAL SOCIETY 75472

DESIGNING A NEW MACHINE

This sketch by the Henry Dreyfuss Agency shows a preliminary stage of John Deere's New Generation tractor design. Note the "V" engine configuration. John Deere extensively tested V-4 and V-6 engines when designing its all-new line, but eventually rejected them in favor of more conventional inline four- and six-cylinder engines. All aspects of design were considered fair game when John Deere created this new look. Even a bright yellow hood and an all-brown color scheme made their way into early concept sketches.

JD C14-1
JR 1-30-59

▲ **This 1959 sketch was drafted with a brush and colorized in the signature John Deere color scheme with black, green, and yellow wash.** Cooper Hewitt, Smithsonian Design Museum / Art Resource, NY

54 · The New Generation

1961 John Deere 4010 SN 1 · Origin: **Moline, Illinois** · Company: **John Deere**

Early in 1953, after years of acrimonious debate, the leaders at John Deere decided to build a tractor engine with more than two cylinders. The two-cylinder had been a mainstay for the line, but everyone from engineering to upper management believed the limits of the technology had been explored and the time had come to change.

The decisions were protected with utmost secrecy, so much so that no records exist of the actual decision to build the new engine (and new tractor).

John Deere was well positioned for the change—it had been the number two tractor builder in the United States for decades, and had a firm grip on that position. The conservative company had been content with that for decades, but that was beginning to change in the 1950s.

For the New Generation tractors to succeed, the new machines needed to make a quantum leap in technology. John Deere's development cycle is legendarily long to this day, and the New Generation would take roughly seven years from concept to introduction. To be certain this could be done without alerting the competition, the work needed to be done in secret.

Secrets are not easily kept by tractor companies. Manufacturing facilities for competing brands are often located in the same town. The red and green combine plants, for example, were nearly back-to-back in Moline, Illinois, and everyone from managers to line

▼ **John Deere's New Generation tractors were a complete redesign and a major departure from the company's staid, two-cylinder heritage. This Model 4010 New Generation is serial number 1000, the first Model 4010 manufactured.**
Keller Collection / Lee Klancher

workers would take lunch, smoke breaks, and reconnaissance missions to peer at the nearby plant and see what was shipping out.

Employees also would switch companies, and every locale in a small town, from the schools to the bars to the softball fields, was a place where "red" employees and "green" employees, would mingle and swap stories. Like any small town, word about interesting developments had a way of spreading like wildfire.

Agricultural sites where the newest machines were tested were also often well known and frequented by lots of different color manufacturers. During harvest or planting season at these locations, company spies would watch for prototype machines in the fields. Actual examples of tractor spy skullduggery included crawling underneath the machines in the wee hours of the morning to scrape off metal samples, disassembling parts of machines hidden under canvas covers, and even management pretending to be dealership owners in order to drive the prototypes.

To prevent this type of corporate espionage with the New Generation, John Deere sequestered a small engineering team in the dark corners of a rented Waterloo grocery store known to its denizens as the "butcher shop." Behind locked doors and blacked-out windows, and at times without heat or air conditioning, the team created the earliest designs for the largest of the all-new John Deere tractors.

The first engine designs were for a V-4 and a V-6, which the sales team desperately wanted. They had been using the fact that their engines were unique as a sales tool since the 1920s, and they wanted to continue to offer that edge. Unfortunately, the V-engines didn't work, and that track was eventually scrapped in favor of a more traditional inline design.

While the engineers were at work, the CEO of John Deere, Charles Wiman, became ill and passed the reins to William Hewitt in 1954. By 1955, Hewitt had issued an edict: "Pass International Harvester." He knew that in order to do so, the company not only needed the engineers to build better product, but it also needed to improve its sales and marketing machine.

The team was given broad parameters developed by upper management, and was instructed to create a completely new machine. The multicylinder engines were the biggest news, but the hydraulic system, transmission, frame, bodywork, hitch, and control systems were all new as well. Nearly every functioning system on the new tractors was redesigned and improved.

▼ The dash on the New Generation tractors was clean and simple. Known as "human factors engineering," machines were designed to improve their physical interaction with the operators. John Deere engineers and HDA worked hard to make the New Generation machines a step up in ergonomics and control-use ease.
KELLER COLLECTION / LEE KLANCHER

The new tractors' appearance was dictated by function, with the most styling attention given to the hood. The curves were carefully designed so that the tractor looked "right" whether it was working at an angle in a field or sitting perfectly level. This subtlety was felt to be a crucial way to distinguish the machine in the marketplace. The carefully designed curves of the hood were complemented by seamless surfaces—every screw head and panel junction possible was concealed.

Although Wiman had famously suggested that the only carryover design element from the previous machines should be the green-and-yellow paint scheme, on the design board nothing was sacred. Prototype drawings depicted yellow tractors with inset stainless-steel panels, and even brown tractors.

Under that seamless hood would sit the star of the new line: an engine with more than two cylinders.

The program had begun in 1953 with the intention of having finished machines ready to roll out in 1958. As the design of the new machines progressed, it became apparent that redesigning every single system properly would take more time. The introduction of the new line was delayed. In the meantime, the 20 series was "redesigned" as the 30 series.

Finally, the new machines were introduced with tremendous fanfare. More than six thousand people were flown to Dallas, Texas, and at noon on August 30, 1960, the new 3010 tractor was unveiled at the Neiman Marcus store in downtown Dallas. Harold Stanley Marcus, the flamboyant owner of the store, was the master of ceremonies at this private unveiling. Guests found a giant gift-wrapped package near the jewelry counter. Tish Hewitt, the wife of John Deere leader Bill Hewitt, cut open the bow and the package to reveal the 3010, with diamonds taped to its flanks and a diamond corset on its muffler. Speeches and other introductions to the new machines lasted all weekend. The entire gala, known as Deere Day in Dallas, was a massive party, capped off with a fireworks display at the Cotton Bowl.

Four models were introduced in Dallas: the 3010 and 4010, as well as the 1010 and 2010. John Deere's New Generation may have arrived glittering with style and jewels, but the machine was conceived in an old grocery store in Iowa.

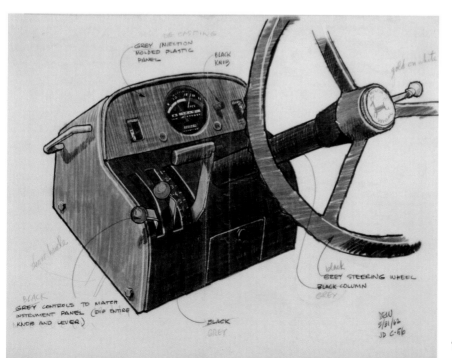

◀ **This sketch by design company Henry Dreyfuss Associates (HDA), dated May 31, 1962, is of a revised dash for later models.**
Cooper Hewitt, Smithsonian Design Museum / Art Resource, NY

THE 1961 HORSEPOWER SHOOTOUT

The John Deere 4010 was a game changer, with cutting-edge looks and state-of-the-art ease of operation. Pretty lines and easy handling only got you so far in this era; farmers were working bigger acreage and in desperate need of additional horsepower. The 4010 diesel offered 10 more ponies than any of the competitive row-crop machines. The 4010 had 37 percent more power than the 560, the model offered by the industry leader, International Harvester.

Game and set.

The competition scrambled to respond, and all the major tractor makers had substantially revised models out between 1963 and 1965. Many of them, like International's brilliant 806, were competitive with the 4010.

John Deere rolled out the revised 4020 in 1963, with 83 horsepower at the drawbar, and took over the number one spot in the agricultural tractor industry the same year.

Game, set, match.

Horsepower Wars 1961						
Manufacturer	Year Introduced	Model	Engine	Price	(Year)	Nebraska Tractor Test Rating
John Deere	1961	4010	380 CID six-cylinder diesel	$5,500	(1963)	72.58 drawbar hp
Oliver	1960	1800A	283 CID Waukesha six-cylinder diesel	NA		62.55 drawbar hp
Allis-Chalmers	1961	D19	Turbocharged 262 CID six-cylinder diesel	$5,300	(1964)	62.05 drawbar hp
J. I. Case	1960	831C	301 CID four-cylinder diesel	$6,000	(1969)	58.3 drawbar hp
International Harvester	1958	560	282 CID six-cylinder diesel	$5,500	(1963)	53.12 drawbar hp

THE HUNTED

1962 John Deere 4010 High-Crop Gas · Origin: **Moline, Illinois** · Company: **John Deere**

The rarest John Deere tractor serial numbers are committed to the memories of the most experienced and dedicated enthusiasts. The sharpest collectors can rattle off the serial numbers of these holy grails—machines so rare that their values are set by the seller rather than the buyer.

A few dedicated individuals—"tractor hunters," if you will—seek out these tractors, knowing that if they can find them, they can charge high finder's fees. They also know who the collectors are that prefer (and can pay for) the rarest machines on the planet. The limited production machines are often oddball variants built for specialty crops—the wide and narrow, hi-crop and orchard—and all of these were used heavily in California, which makes the state a favorite hunting ground.

A tractor hunter from Iowa, we'll call him Dan, has found a number of the rarest tractors in existence. One day, Dan called Wisconsin collector Walter Keller, who owns a number of very rare John Deere tractors.

"What is the serial number of that gas 4010 tractor?" Dan asked.

Walter knew there was only one gasoline-powered 4010 built. And Dan knew that Walter knew.

Walter bought it on the spot, and had the money (in cash) sent to Dan.

The tractor is one of the most unique John Deeres in existence. It was fitted with a gasoline engine at the factory and sent to Appleton, Wisconsin. The machine was returned to John Deere, retrofitted with a diesel engine, and then sent out to California, where it worked on a commercial farm, and was branded with the number sixteen as seen in the photograph.

Toy manufacturer Ertl used this very machine to create a scale-model tractor, which originally had a gasoline engine. Ertl received so many complaints from collectors who were absolutely positive a gas model was never built, they changed it to a diesel. The original scale model with the gasoline engine is now highly collectible.

▶ **This tractor—serial number 23T 36420—was originally built as the only gas 4010 high crop. It was later returned and a new diesel engine was put in.**
Keller Collection /
Lee Klancher

55 · Taming the Wilds of Suburbia

1961 Cub Cadet · Origin: **Chicago, Illinois** · Company: **International Harvester Company**

7 and 10 hp tractors for the farm?

Sure. One of every four International Cub Cadets is bought by a farmer like yourself. Used in the same way as by suburban estate owners around their grounds, lawn or garden.

Mowing, for instance. Or mulching leaves. An acre or more an hour. Gentle on the grass. And of course, a real miser with fuel.

Direct drive (no belt) power train is the same

as in bigger IH Cub® tractors. In fact, the Cub Cadet® warranty is the same as for the biggest IH tractors. Overall, Cadets are about 100 pounds stronger than other 7 and 10 hp home tractors.

Arrange a demonstration soon through your IH dealer. See for yourself the Cadet's no toy. It's a small but gutsy IH tractor! International Harvester Company, Chicago, Illinois 60601.

International Harvester—the people who bring you the machines that work

A tractor to cut the grass! This brilliant idea hit the industry with a burst of energy in the late 1950s. One of the first was the 1959 David Bradley Suburban tractor (sold through Sears), though Wheel Horse had lawn tractors out a few years before that. Not surprisingly, shortly after these little tractors hit the market, the large tractor makers lit fires under their engineering teams to respond in kind.

International Harvester hit the market with a machine it initially called the "Cubette." The rig was powered by a Kohler 7-horsepower engine, featured a full line of implements, and experimental Cubettes were frolicking in the dirt by summer 1960.

The name was changed to the Cub Cadet, and the production model was introduced in January 1961. Demand for the machine exceeded expectations, and International sold about 63,000 units by 1963.

The Cub Cadet line remained a strong addition for International Harvester until the company ran into financial troubles in the early 1980s, and it sold the line to Modern Tool and Die Company in 1981.

▲ At the 1966 Farm Progress Show in Farmer City, Iowa, three custom-painted Cub Cadets took center stage with vaudeville actor George Mann playing the character "Fisbee—the hapless animal trainer."
Wisconsin Historical Society 26992

56 · No Cigars?

1961 Allis-Chalmers B-1 · Origin: **West Allis, Wisconsin** · Company: **Allis-Chalmers Company**

Allis-Chalmers joined the lawn tractor fray in 1961 with this Simplicity-sourced Briggs and Stratton–powered machine. The marketing department clearly had a sense of humor.

WHATDYAMEAN...NO CIGARS?

◀ **The 1961 Allis-Chalmers B-1 is born.**
OCTANE PRESS COLLECTION

57 · Matching Drapes

1969 John Deere Model 112 · Origin: **Moline, Illinois** · Company: **John Deere**

John Deere's first garden tractor, Model 110, was introduced in 1963. This is the Model 112, which famously were offered in four different colors in 1969: Sunset Orange, Spruce Blue, Patio Red, and April Yellow. These sold poorly, and production of these colored machines ended in 1971. They are now known as the "Patio Series" lawn tractors.

▶ **The Sunset Orange version of the John Deere Model 112 "Patio Series."**
KELLER COLLECTION / LEE KLANCHER

58 · Engineering Can't Always Save the Day

Oliver OC-4 · Origin: **Charles City, Iowa** · Company: **Oliver Farm Equipment Company**

The Oliver OC-4 was proof that engineering was not everything. Originally created by Cletrac, these crawlers were technically sound machines produced by a fiscally unsound organization. The machine was a solid crawler that sold and performed well for Oliver. The machine's merit alone was not enough to keep it alive however. White purchased Oliver in 1960 and Cletrac in 1961. White had a poor dealership distribution and low market share for the innovative crawlers, and production of the Oliver OC-4 was discontinued after 1965.

◀ **This experimental model photographed on April 22, 1960, would have become the OC-5. White killed the line, and this crawler was never produced.** FLOYD COUNTY HISTORICAL SOCIETY

59 · Beauty Is Skin Deep

1963 John Deere Model 2010C · Origin: **Moline, Illinois** · Company: **John Deere**

▶ **This is a 1963 John Deere 2010 diesel crawler formerly owned by Washington State University. Only 159 2010 crawlers were built with green paint. Ertl Company based their model off this example.** KELLER COLLECTION / LEE KLANCHER

The push for a light, agricultural crawler of roughly 50 horsepower was on in the early 1960s, and the two leading manufacturers (red and green) both produced weak entries into this market. The New Generation item was this 2010 crawler, a beautiful machine flawed by a weak driveline.

60 · Power Isn't Everything

International BTD-6 · Origin: **Doncaster, England** · Company: **International Harvester Company**

In the early 1960s, power was all the rage. International loaded their light crawler with all they could find in the 55-horsepower, six-cylinder DT282 engine that appeared on the Series 62 versions. The engine could (and did) easily overpower the driveline, and the model was not well received. Note that the American version was built in Melrose Park, Illinois, while the British version was built in the UK.

◄ **This BTD-6 and Drott buckethead are at work on St. Paul's Cathedral in London in 1963.** WISCONSIN HISTORICAL SOCIETY 6674

61 · The People's Tractor

1960 Porsche Junior · Origin: **Dusseldorf, Germany** · Company: **Mannesmann AG**

▲ This 1960 model of the Porsche Junior was imported from Germany by one of the members of the DuPont family. The machine is powered by a Porsche single-cylinder diesel engine with a hydraulic coupling between the engine and transmission.

GRUNNAH COLLECTION / LEE KLANCHER

Dr. Ferdinand Porsche created the vision for a tractor in the 1930s during the time that he was developing the people's car, which eventually became the Volkswagen Beetle. His early design featured a simple, air-cooled engine, progressive styling, and a fluid coupler (instead of a clutch) between the engine and transmission. Four models were created with one-, two-, three-, and four-cylinder air-cooled diesel engines.

After World War II, Porsche was not allowed to produce the machine, and production was licensed to a German company, Allgaier Werke GmbH, and an Austrian company, Hofherr-Schrantz. In 1956, Mannesmann AG (a growing German conglomerate) purchased the rights and began building Porsche tractors in an old Zeppelin factory in Germany. The company had great success with the machine, selling more than 125,000 units between 1956 and 1964. Roughly 1,000 of those were imported to the United States at the time, with many more coming over the pond in the modern day as collectors purchased them.

The company lost interest in the agricultural sector, and stopped producing the tractor. Mannesmann went on to build a powerful telecommunications network, and was purchased by Vodafone in 2000.

62 · Stylish Endings

1958 Cockshutt Model 560 · Origin: **Brantford, Ontario** · Company: **Cockshutt Farm Equipment Co.**

Cockshutt is the tractor industry's chameleon, a company that changed colors and brands seemingly at will. The Canadian agricultural implement company was founded in the late 1800s to build plows. Cockshutt handled Canadian distribution of Allis-Chalmers tractors in the 1920s, and Oliver tractors in the 1930s. Some of the Olivers were branded Cockshutt and painted red. Cockshutt built its own tractors during World War II, and introduced a line of new machines in 1958.

The new 500 series line of Cockshutts was designed by none other than famed industrial designer Raymond Loewy, who had done among many other things the Farmall Letter Series introduced in 1939 (see tractor number 28 for more on this). He gave the 500 series stylish designs, and a new color scheme to boot.

Just as the company was readying its new line for the market, Cockshutt was taken over by outside interests and the farm equipment division was sold to the White Motor Company.

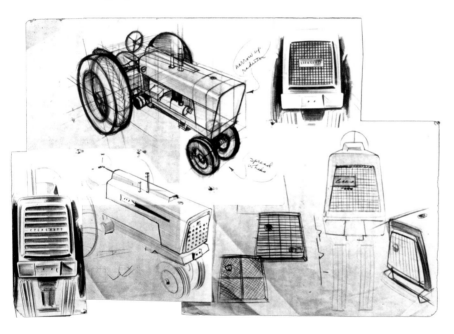

◀ **Loewy sketch of proposed Cockshutt 500 series tractor design.**
RAYMOND LOEWY ™ BY GOOD DESIGN, LLC. RAYMONDLOEWY.COM

◀ **The Cockshutt 560 was built from 1958 to 1961.**
RAYMOND LOEWY ™ BY GOOD DESIGN, LLC. RAYMONDLOEWY.COM

63 · Horsepower and Hubris

1961 International 4300 · Origin: **Chicago, Illinois** · Company: **IHC / Hough Industries**

▼ **The IH 4300 was produced from 1961 to 1965. Power came from an IH 817-cubic-inch DT817 turbocharged diesel engine rated for 300 crankshaft horsepower. The tractor weighed 15 tons, so power to weight wasn't ideal, and it lacked a three-point hitch.**
Mez Collection / Lee Klancher

In January 1959, up-and-coming IHC executive Brooks McCormick visited the company office in Portland, Oregon, to inquire just precisely what was happening with sales in the region. The Wagner four-wheel-drives were selling well in the area, and Brooks wanted to know how and why. Compounding this was the fact that IHC had gotten wind of John Deere's soon-to-be-introduced high-horsepower four-wheel-drive 8010.

Memos started flying as IHC management switched to high alert: they needed a four-wheel-drive, and they needed it yesterday.

IHC subsidiary Hough, a construction company and fully owned subsidiary, had been working on a solution. They presented a lavish proposal for the 4WD-1, a big tractor that was nearly ready for production. IHC bit hard, wrote a big check, and started development. Early machines had insufficient power, so a 4WD-3 was developed using a 300-horsepower IHC engine and making the world's most powerful farm tractor.

By the time the tractor was ready for production in 1961, the company's enthusiasm for the machine had dimmed, and the introduction was soft-pedaled with minimal promotion and sales effort. The big machine was too heavy and clumsy for field work, and was bigger than the market demanded at the time. Fewer than fifty machines were sold, and IHC would have to find another solution for the four-wheel-drive market.

64 · West Coast Power Trip

1961 Wagner TRS14 · Origin: **Portland, Oregon** · Company: **FWD Wagner, Inc.**

In 1961, four-wheel-drive truck manufacturer FWD purchased Wagner and created FWD Wagner. The Wagner model designation changed from "TR" to "WA" after the purchase. It had a wide variety of four-wheel-drive tractors ranging from 100 to 280 horsepower. These machines were the class of the four-wheel-drive market at the time, and were overbuilt, well-supported machines.

▶ **Wagner's high-horsepower line of four-wheel-drives was the state of the art in the early 1960s.** Wisconsin Historical Society 129044

65 · Forced Air

1961 Allis-Chalmers D19 · Origin: **West Allis, Wisconsin** · Company: **Allis-Chalmers Company**

The largest portion of the 1960s tractor market was occupied by row-crop tractors, and the midsize farms that used them were also looking for boosted power. The West Allis crew hit the market with the diesel D19, which featured the row-crop market's first factory standard turbocharger. The D21 would follow in 1963, and it would be the first Allis-Chalmers to break the 100 horsepower barrier.

◀ **The Allis-Chalmers D19 was produced from 1961 to 1964.** In addition to the turbocharged diesel engine, the model could include gasoline- and LP-burning power plants. Gary Nelson

66 · The Hired Man

Oliver 1600 Hi-Crop LP · Origin: **Charles City, Iowa** · Company: **Oliver**

▼ **The Oliver 1600 was introduced in 1962 and produced until 1964. Even a boost in displacement in 1963 didn't give it enough power for its class.**
Lulich Collection / Lee Klancher

The lively engineering crew housed at Charles City, Iowa, had a lot of tricks up its sleeves in the 1960s. Well aware that labor was hard to come by on the farms of the day, the team designed the 1800 / 1900 tractors introduced in 1960 to provide enough power and convenience to replace a hired man.

Led by Herbert T. Morrell, the Charles City engineering team continued to roll out interesting innovations throughout the decade, with nifty items like cast iron grilles designed to help balance weight, a hydra-power drive transmission, tilting and telescoping steering wheel, wheel guard fuel tanks, and giant Terra tires with such light ground pressure that a salesperson drove a tractor equipped with the big wheels over chicken eggs at the Farm Progress Show. He didn't break a single shell.

Despite their efforts, Oliver's fate was sealed when it lost control due to a corporate takeover and was purchased by White Motors in November, 1960. The brand lived on, but the eggs would prove to be in the wrong basket.

67 · The Outlier

1965 Minneapolis-Moline G708 · Origin: **Hopkins, Minnesota** · Company: **White Motor Company**

The Minneapolis-Moline G series of tractors came out in 1953, and ranged from the G-704 to the G-1350. The 1350 was one of the last, and featured 142 PTO horsepower. In 1963, Minneapolis-Moline became another of the brands absorbed by the White Motor Company.

▶ **The G708 was powered by a 504-cubic-inch six-cylinder, and put out more than 100 horsepower at the crankshaft. Produced only in 1965, this was a very rare machine.** Lee Klancher

68 · A Global Perspective

1971 Massey Ferguson 1100 · Origin: **Brantford, Ontario** · Company: **Massey Ferguson Limited**

Massey Ferguson survived the consolidation surge of the 1960s by being the shark rather than the minnow. The Canadian-based firm used acquisition to grow, and it purchased three major makers in less than a decade. This included Landini, an Italian tractor maker; a portion of Ebro (Motor Ibérica), a Spanish tractor and car maker; and Perkins Engines, which built diesel engines. Massey Ferguson used these acquisitions to help build one of the best worldwide sales networks in the business, and even when times were tough, it moved more machinery than all but the very top agricultural equipment companies.

▶ **The Massey Ferguson 1000 series tractors were generally solid machines. The 1100 was produced from 1964 to 1971 and produced 94 PTO horsepower.** Winslow Collection / Lee Klancher

69 · The Powerhouse

1963 John Deere 5010 · Origin: **Moline, Illinois** · Company: **John Deere**

The John Deere 5010 was tested at the Nebraska Tractor Test Laboratory in October 1962, where the 531-cubic-inch inline six-cylinder diesel engine was found to produce 121.12 PTO horsepower and (drum roll) 105.92 horsepower at the drawbar.

APPLIED GRAPHICS & COLOR BAR

SEPARATE STAMPED PART ATTACHED TO HOOD

◀ **This Henry Dreyfuss sketch was drafted in 1961 and served as a preliminary design for the Model 5010.** COOPER HEWITT, SMITHSONIAN DESIGN MUSEUM / ART RESOURCE, NY

▲ **The Model 5010 was produced from 1963 to 1965. This unit is serial number 23T 01000, the first production model built.** KELLER COLLECTION / LEE KLANCHER

70 · Flying High

1964 John Deere 4020 · Origin: **Moline, Illinois** · Company: **John Deere**

When John Deere released the New Generation in 1960, it shocked the industry and dominated the segment. They sold more than 57,000 4010s alone. In fall 1963, the men in green played another card by introducing a revised line. The 4020 featured more displacement and power—the diesel version produced 94 PTO horsepower when tested at Nebraska.

The power was good but not class leading. The ace in the hole for the 4020 was the powershift transmission, which allowed the operator to shift on the fly. Although the competition had a variety of torque amplification systems that generally allowed the operator to select between two speeds while moving, Deere's optional powershift transmission could move through the entire eight-speed forward or four-speed reverse range while under load. Coupled with refinements to fix minor issues and upgrade others, the new 4020 was the standout model in the segment.

The 4020 was also a critical machine in farm safety, as the roll-over protection system (ROPS) was developed on that model. The system was a heavily built steel bar crossbar and risers installed on open platform tractors designed to prevent the tractor from squashing the operator as it rolled. Deere released the feature in 1966, and it eventually became an industry standard for open platform tractors. It saved the lives of thousands of farmers.

The 4020 helped John Deere take over the number one US sales position in agricultural equipment in 1963, a spot it would hold until modern times. John Deere also happened to be the most profitable agricultural maker in the industry— a factor that would be critical in the hard times to come.

▼ **This 4020 is the first powershift model 4020 off the line, and the second 4020 built.**
PURINTON COLLECTION / LEE KLANCHER

71 · Fighting Back

1965 International 1206 · Origin: Chicago, Illinois · Company: International Harvester Company

The 1206 was International's flagship in 1965. Equipped with a turbocharger designed for the model by IH subsidiary Solar Turbines, the 1206 was a heavily beefed-up variation on the 806. With improvements in the engine, final drive, and more, the 1206 proved to be a strong entry in the hotly contested high-horsepower two-wheel-drive segment.

◀ **The 1206 is an all-time favorite IH tractor, and the first two-wheel-drive International to produce more than 100 horsepower.** MARK JENSON

HORSEPOWER WARS 1965

A fierce fight for power continued in the mid-1960s. The 4020 and the 806 were both introduced in fall 1963, and were surprisingly comparable in power. The John Deere had the edge in transmissions with a powershift unit. The Oliver 1950 came out a year later with a turbocharged two-stroke engine topped in output only by the 5010.

The big wheatland tractor war got interesting in 1965, with the 1206 making a run but outgunned by both John Deere and the surprisingly strong Allis-Chalmers D21 Series II. In the end, the 5020's 141 horsepower made it the king of the two-wheel-drive world in the mid-1960s.

Manufacturer	Year Introduced	Model	Engine	Price	(Year)	Nebraska Tractor Test Rating
John Deere	1965	5020	531 CID six-cylinder diesel	$15,000	(1972)	141 PTO hp
Allis-Chalmers	1965	D21 Series II	426 CID six-cylinder turbocharged diesel	$5,700	(1964)	127 PTO hp
John Deere	1963	5010	531 CID six-cylinder diesel	$11,000	(1965)	121 PTO hp
International Harvester	1965	1206	361 CID six-cylinder turbocharged diesel	$9,450	(1967)	112 PTO hp
Oliver	1964	1950	212 CID four-cylinder two-cycle turbocharged diesel	$12,000	(1974)	105.7 PTO hp
Minneapolis-Moline	1965	G708	506 CID six-cylinder diesel	$9,000	(1965)	101 PTO hp (claimed)
John Deere	1964	4020	404 CID six-cylinder diesel	$10,345	(1972)	95.83 PTO hp
International Harvester	1964	806	361 CID six-cylinder diesel	$6,800	(1967)	94.93 PTO hp

72 · The Opportunity

1964 J. I. Case 1200 Traction King · Origin: **Racine, Wisconsin** · Company: **J. I. Case**

In 1963, the J. I. Case engineering team started developing a midsize, modestly priced, 100-horsepower four-wheel-drive based on its W-12 Terraload'r. A key feature for the model was crab steering, which allowed all four wheels to turn. Full production of the new 1200TK farm machine began in 1964 at the Racine, Wisconsin, plant with later models built in Rockford, Illinois. The 1200TK was a hit, with more than 2,041 of these built. More important, J. I. Case had found a niche with a value-priced, crab-steering, four-wheel-drive machine.

▲ The 1200 Traction King was produced from 1964 to 1969, and was powered by a Case A451D turbocharged engine good for 119.9 PTO horsepower.
CASE IH

FERTILE FIELD FUNDING

Throughout the 1950s, J. I. Case suffered tumultuous times, due to several changes in management and structure along with nearly constant multimillion-dollar annual losses. By April 2, 1964, the *New York Times* reported J. I. Case had in excess of $256 million in debt, and had registered only $130 million in gross revenue in 1963.

The Kern County Land Company headquartered in Bakersfield, California, had extensive agricultural land and oil field holdings. It had an eye on J. I. Case

and began angling for a purchase. As reported in the *New York Times* on April 21, 1964, the Kern County Land Company bought a controlling interest in J. I. Case for a purchase price of about $30 million (which included a $46 million tax credit).

Pay attention—this purchase of a tractor industry company was by a land-holding company with both oil and agricultural interests, a trend that would play a significant role in which companies would survive to modern times.

73 · When Doug Met Earl

1965 Steiger 2200 · Origin: **Fargo, North Dakota** · Company: **Steiger Manufacturing, Inc.**

▲ **The Steiger 2200 is part of the "barn-built" machines built in the Steiger family's barn converted to a shop near Red Lake Falls, Minnesota.** Oschner Collection / Lee Klancher

In 1963, the Steiger family displayed one of their prototype machines at a farm show in Crookston, Minnesota. Local dealer / owner Earl Christianson saw the nicely designed green four-wheel-drive and smelled an opportunity. He went to meet Douglass Steiger, and told him if he would build more of the tractors, Earl would sell them. Doug made one of the best decisions of his life when he agreed. Earl's gift for sales and promotion raised Steiger to the next level, and the machines that resulted from his and Doug's deal were the Steiger 1700 and 2200, two all-new models that featured intelligent design, quality components, and the ruggedly overbuilt construction that would define Steiger as the "Cadillac" of the four-wheel-drive market.

74 · A Challenger

1966 Versatile G-100 · Origin: **Winnipeg, Manitoba** · Company: **Versatile Manufacturing, Ltd.**

▲ **A Versatile G-100 with a Chrysler V-8 gasoline engine was built in 1966 and 1967. Versatile also offered the diesel-engined D-100 powered by a 128-horsepower Ford 363.** New Holland

A former tool designer at Massey-Harris, Peter Pakosh began building and selling a portable grain auger out of his home in Toronto in 1945. He partnered with brother-in-law Roy Robinson and the pair created a field sprayer and a harrow drawbar. They moved to Winnipeg in 1951, and grew the company with more implements.

In 1963, they created Versatile, and in 1966, introduced small, articulated four-wheel-drive tractors with their own chassis. Their strategy was to enter the market with a high-value, low-cost, four-wheel-drive machine. Versatile found their niche in North America and Canada, and it all began in the backyard of an innovative tool designer.

75 · Acquire and Conquer

1968 John Deere WA-14 · Origin: **Portland, Oregon** · Company: **FWD Wagner**

John Deere's 8010 / 8020 four-wheel-drive was not successful (to say the least), and the company needed a four-wheel-drive offering while it engineered a new machine. On December 31, 1968, John Deere contracted with Wagner to paint their WA-14 and WA-17 machines green and yellow and badge them as John Deere machines. The deal was structured in such a way that the fate of the Wagner tractor depended entirely on John Deere. If it ended, Wagner was prohibited from building a competitive machine five years after the deal ended. John Deere sold only twenty-three WA-14s and twenty-eight WA-17s. Production was stopped in 1970, and Wagner tractors came to an end.

▲ **These Wagner-built John Deere tractors were powered by a Cummins NT855C1 rated for 280 horsepower.** PETER SIMPSON COLLECTION

76 · Think Tank Dreams

1968 International TXA-62 Concept · Origin: **Chicago, Illinois** · Company: **International Harvester Company**

While people such as Douglass Steiger and Peter Pakosh built innovation in their backyards, engineers at the major manufacturers used drawing boards and imagination to dream up ways to supply the farmer with more power, traction, and efficiency. Keith Burnham's drawings of this cab-forward design is one of the most interesting proposed—creating a single machine to handle all farm chores with removable equipment, as well as putting two operators up front where it was cooler and cleaner. Drawings and models were made into the early 1970s, but the TXA-62 never made it past the model room at IH.

▶ **This drawing from November 13, 1967, shows the cab forward, twin-seat concept. The open back would have allowed mounting power units such as a powered combine, nearly unlimited room for large fuel tanks, and improved serviceability.** KEITH BURNHAM COLLECTION

FARMALL Tractor
2 Man Cab Forward Concept

77 · Moving . . . Up?

1969 Steiger Tiger · Origin: **Fargo, North Dakota** · Company: **Steiger Manufacturing, Inc.**

▲ **This Steiger Tiger is an early prototype that was built out in the Steiger family barn and (probably) completed in the new building in Fargo, North Dakota.**
Bonanzaville
Collection /
Lee Klancher

On January 22, 1969, Steiger took a leap forward with a major investment that created a new company and moved the headquarters off the farm in Minnesota to a dilapidated warehouse rented in Fargo, North Dakota. A new management group included CEO Robert Kelly, who slept during the day and worked at night, and a chief engineer who was hard to find at best. While this move was a sidestep at best, the deal also brought in some bright financial minds who would work with the Steiger family to drive the company to become an industry leader in four-wheel-drive tractors.

78 · What Could Have Been

1969 Minneapolis-Moline A4T-1900 Experimental · Origin: **Hopkins, Minnesota** · Company: **Minneapolis-Moline**

By Sarah Muirhead, *Max's Tractor Shed*

Only one experimental A4T-1900 tractor was ever built. Its lead engineer was Mike Verhulst of Ottumwa, Iowa, who, at the time the A4T-1900 was in concept stage, was a mere twenty-three years old.

Minneapolis-Moline entered the four-wheel-drive market in 1969 with its A4T-1400 4WD articulated tractor, followed by several versions of the A4T-1600. The plan from there was to change the A4T-1600 to A4T-1700 with a new transmission and a non-turbocharged D585 engine. A larger model with a turbocharged engine and the new transmission would be added from there, and named the A4T-1900. This model would be diesel-powered only.

Plans changed in the fall of 1971 when White Motor Company officials decided to close the Lake Street plant and eliminate the Minneapolis-Moline sales and manufacturing operations. All tractor manufacturing was moved to Charles City, Iowa, and marketing was incorporated into the Oliver organization. Work on the A4T-1900 was halted.

The only experimental A4T-1900 that was built was converted and used for some time as a load test mule on the test track at Charles City. It was purchased, restored, and at the Half Century of Farm Progress Show was reunited with engineer Mike Verhulst, who drove it on the show grounds.

▼ While Minneapolis-Moline did have several A4T four-wheel-drives in production, this A4T-1900 is an experimental model and the only one that exists.

MAX'S TRACTOR SHED / SARAH MUIRHEAD

79 · The Green Turbo Diesel

1970 John Deere 4520 · Origin: **Moline, Illinois** · Company: **John Deere**

By Jim Allen

▲ Despite having the same engine, the standard transmission and powershift 4520 tractors had different performance. PTO power was a fraction less in the powershift test, and it developed slightly less drawbar horsepower (108 hp vs 111 hp). Conversely, the powershift developed dramatically more drawbar pull (10,079 lbs versus 8,773 lbs), while delivering better fuel economy. This 1970 Model 4520 was equipped with rare front-wheel assist and a powershift transmission. The books say 7,894 4520 tractors were sold. KELLER COLLECTION / LEE KLANCHER

In 1961, Allis-Chalmers was the first company to produce a turbo diesel tractor, the D19. Though the result was appreciated and admired, the other tractor manufacturers took their time developing their own turbo diesel tractors. When needing more power, the manufacturers, and perhaps the American farmer, still had a preference for big-displacement, slow-turning, naturally aspirated diesels. Consider, too, that the old diesel designs still in use in the 1960s did not lend themselves to turbocharging. That wasn't true of John Deere.

John Deere used 381- and 531-cubic-inch engines in the 4010 and 5010. When the 4010 became the 4020 in 1963, the 381-cubic-inch six-cylinder engine was given a displacement boost to 404 cubic inches. The stage was then set for a new tractor.

John Deere was a company that generally did not allow itself to be rushed into putting new developments on the market before they were ready. In fact, John Deere had benefited when other companies had done so. Although some may have clucked a little about John Deere being a little behind the others in offering a turbo diesel tractor, the folks at John Deere were confident they had not stepped on their crank when they finally did.

The 4520 debuted for the 1969 model year as the company's first turbocharged diesel tractor. It fit between the very popular naturally aspirated 4020 and the much larger 5020, with its 531-cubic-inch six-cylinder engine. Some have called the 4520 a "4020 on steroids," but that's not precisely true. It used a 404-cubic-inch engine that was based on the same NA engine used in the 4020, but the power train and the tractor were larger and heavier. A more accurate description might be to call it a downrated 5020 rather than an uprated 4020. Unlike some of the other turbocharged tractors on the market, the 4520 wasn't just a tractor to which a turbo was added. The entire package had been built around the extra power, and John Deere used the ad line "Turbo-Built."

By 1969, the 404 six-cylinder had become one of John Deere's mainstays. It was, and is, a highly effective power plant that many think was seminal in cementing the company's success in the 1960s and 1970s. The turbo 404 boosted the PTO power output to a modest 122. The compression ratio was lower than the NA version (15.7:1 vs 16.5) with stronger pistons that used keystone rings and had uprated piston oil cooling jets. The turbo engine had a beefier new block with more main bearing support and improved oil flow.

The most whiz-bang part of the new 4520 was the air filtration system. Designed to both clean the inlet air in a superior fashion *and* offer a long maintenance interval, the air cleaner had twelve venturis that removed dirt backed up. Along with this, a special muffler was designed to siphon off roughly 90 percent of the incoming dirt and blow it out the stack. Pretty clever huh? Unfortunately, this turned out to be a place where John Deere might have tripped over its crank. It didn't take long for problems to start—dirt began being blown into the engine, and in some cases even set the air filters on fire. John Deere had to pay for upgrades to customers' tractors even after the normal warranty had expired. Whoops! With an upgrade, the 4520 became a tractor to admire.

The 4520 was available with either the standard Synchro-Shift, or with the Power-Shift that offered clutchless shift on the fly. It was one of the first tractors with a ROPS (Rollover Protection System), which John Deere called "Roll-Gard." A heated and air conditioned cab was available as well as a fiberglass canopy that attached to the ROPS. Later in the run, an FWA system was offered as an option. Some of the last 4520s built in 1972 have been seen with the Sound-Gard cab that was one of the major contributors of the hoopla to John Deere's "Generation II" program.

The 4520 lasted until 1970 when it was replaced by the 4620, which was very similar but aftercooled and made 135 horsepower. The even more powerful 4630 of 1973 made 150 horsepower with an intercooled 404 turbo engine. The 1973 model year was John Deere's second new beginning—akin to the model's 1960 rebirth—and the company debuted a great deal of new technology. Many say that the 1973 Generation II era marks the beginning of John Deere's decades-long domination of the agriculture market. The 4520 was a major stepping stone into that era.

THE DAWN OF POWER FARMING
1970–1979

As the farm modernized, demand exploded for machines capable of working large plots of ground with increasingly sophisticated (not to mention power-hungry) implements. These simple, powerful machines fit the bill. LEE KLANCHER

80 · If You Build It . . .

1970 Big Bud HN320 · Origin: **Havre, Montana** · Company: **Northern Manufacturing Company**

When the John Deere company took the exclusive rights to Wagner on New Year's Eve 1968, the dealers who sold Wagners (and didn't sell John Deere) were left out in the cold. In 1961, Willie Hensler of Havre, Montana, had started their dealership specifically to sell Wagners.

When the John Deere deal went down, Hensler and his service manager, Bud Nelson, focused on repairing and rebuilding old Wagner tractors. They repainted one of those rebuilds white and dubbed it a "Big Bud" tractor. They decided to form the Northern Manufacturing Company to build all-new Big Bud tractors.

The first Big Bud model, the HN-250, sourced a lot of parts used on the Wagner machines, and it looked a lot like a Wagner. The machine featured several innovations: the hood and cab tilted up for engine, transmission, and cooling access; the engine, radiator, and transmission could slide out on a pan, making service a snap. The Cummins NT855 was turbocharged, intercooled, and rated for 250 horsepower.

The rugged, powerful, and relatively simple Big Bud sold well, and the brand became one of the iconic tractor makers of the 1970s and 1980s.

▼ The Big Bud HN-320 was the company's second model. "HN" are the initials for the last names of founders Willie Hensler and Bud Nelson.
MICHAEL HOOD

81 · Homegrown

1972 John Deere 7520 · Origin: **Moline, Illinois** · Company: **John Deere**

The 1960s four-wheel-drive market was a puzzle the industry leaders had a hard time solving, and things didn't improve all that much in the early 1970s. John Deere dropped the Wagner-source WA-14 and WA-17—which sold fewer than 100 units anyway—and focused its efforts on its in-house machine, the new-for-1971 7020. The tractor's 146 PTO horsepower engine was the first tested at Nebraska with an intercooler, which cooled the air going into the turbocharger, making the air denser and oxygen-rich, hence boosting horsepower. Criticized for being complex to service and somewhat fragile, the 7020 was nevertheless a solid seller in the four-wheel-drive class. The upgrade from the 7020 to the 7520 came only a year later, with a larger displacement engine giving the later model a welcomed power boost.

▶ **The 531-cubic-inch six-cylinder turbo diesel engine in the 7520 produced 175.8 PTO horsepower when tested at Nebraska in June 1972.** Geoff Wing

82 · New Style

1972 J. I. Case 1370 · Origin: **Racine, Wisconsin** · Company: **J. I. Case Company**

Fueled with investment from its new owners from Texas, J. I. Case entered the 1970s with a new, high-styled line of machines, the 70 series tractors. The line's 970 was the first to be sound-tested at the Nebraska Tractor Tests, a procedure that took place on twenty-nine tractors in 1970 and later became standard practice. Sound was a growing concern for the farmer, and the cabs of the era were not quiet places.

▶ **The largest row-crop machine in the J. I. Case lineup was the 1370, which was produced from 1971 to 1978. The turbocharged diesel 504-cubic-inch engine put out 155 PTO horsepower when tested at Nebraska.** Case IH

83 · Network Power

1971 International 1066 · Origin: **Chicago, Illinois** · Company: **International Harvester Company**

The International 66 series machines proved the power of good marketing and an influential sales network. Introduced in 1971, the series was heavy on graphics and small refinements, and light on the high-technology features demanded by the market—notably a powershift transmission. Sales were robust anyway, and a 66 series tractor carried the honor of being IHC's five millionth tractor built. Bear in mind this number factored in production from all of IHC's factories in France, Britain, Australia, Germany, and Asia.

As chronicled in a video by Greg "Machinery Pete" Peterson, the five millionth 1066 was auctioned off on September 26, 1976, when the 86 series was released. A group of Montana dealers led by Art Weaver won the bid, paying $40,086.86 for the tractor. The funds were donated to Montana State University for a study to improve tractor operating efficiency.

The dealers shared ownership of the 1066 and used the historic piece in their showrooms, at shows, and during events. As of May 2016, the tractor had only fifty-three hours and was housed at the Museum of the Northern Great Plains in Fort Benton, Montana.

▶ **The International 1066 debuted in 1971, and was produced through the 1976 model year (which was also the only year that the tractor had a distinctive "black stripe" graphic). This International 1066 was dubbed by IHC the five millionth tractor and was produced at the Farmall Plant in Rock Island, Illinois, on February 1, 1974. It is shown with local model Valerie Robb, and the vice president of marketing, Stanley Lancaster.** WISCONSIN HISTORICAL SOCIETY 11275

84 · The Blue Beast

1972 Ford 9600 · Origin: **Dearborn, Michigan** · Company: **Ford Motor Company**

For 1972, Ford stepped up its game and offered the 9600, a high-horsepower two-wheel-drive equipped with a cab that had optional (if dubiously reliable) air conditioning. The 9600 was the largest in the line, and sold for $17,300 in 1976.

▶ The Ford 9600 was produced from 1972 to 1976. The 401-cubic-inch turbocharged diesel produced 135 PTO horsepower when tested at Nebraska. The Fordson next to the 9600 is Ford's first tractor, one of the most prolific tractors in history.
New Holland

85 · The Machines from Minsk

1973 Belarus MTZ 80 · Origin: **Minsk, Belarus** · Company: **MTZ**

By Martin Rickatson

The largest tractor manufacturing plant in the former USSR was located at Minsk, in the then–Russian satellite state of Belarus, where it remains today. Products were initially exported under the Minsk Tractor Works (Zavod) or MTZ brand, but later labeled simply as "Belarus." The 90 engine horsepower MTZ 80 included features such as a sprung front axle and air braking.

◀ Built in the Minsk Tractor Works (MTZ), the Belarus MTZ 80 design incorporated a number of advanced features, including a spring-suspended front axle and air brakes.
MTZ Belarus

86 · The Next Generation

1973 John Deere 4630 · Origin: **Moline, Illinois** · Company: **John Deere**

In August 1972, John Deere introduced (as 1973 models) its Generation II tractors, a revised line-up that featured new styling, improved cabs, and more horsepower. These were great machines at the time, and the leap forward was not as dramatic as the New Generation in 1960 or the "Iron Horses" in 1979.

▶ **The John Deere 4630 was produced from 1973 until 1977. The tried-and-true 404-cubic-inch engine was aftercooled and turbocharged, and produced 150.6 PTO horsepower when tested at Nebraska in October 1972.**

JIM ALLEN COLLECTION

87 · New Life

1973 International 4366 · Origin: **Fargo, North Dakota** · Company: **Steiger Manufacturing, Inc.**

In 1971, the Steiger Tractor Company was barely hanging on in the market, and significant investments in cash in 1969 and again in 1970 kept it alive only by the skin of its teeth. The second influx of cash came from the Melroe family, who had founded Bobcat and sold much of it. The Melroes brought cash to the table—more importantly, they brought in new management.

The talented and experienced crew would make the most important move for the company since the Steigers built the first model in their barn in 1958.

What they did was shop around Steiger as a contract builder, producing four-wheel-drive machines in any color. They would land contracts with Allis-Chalmers and International early on, and the IH contract was a huge one that saved the company.

IH agreed to a three-year deal to make 1,500 machines, allowing Steiger to build a brand-new factory in Fargo to ramp up production. That factory stands today, still building four-wheel-drive tractors, in no small part to the deal that made the 4366.

▼ The 4366 used a Steiger chassis with an International engine, driveline, paint, and decals. The first one was built by young engineer Paul Nystuen, who teamed up with master fabricators Jerry Joubert and Al Lieberg in a secret garage in Fargo in 1972. This is that team the day they finished the first prototype. PAUL NYSTUEN IMAGE COLLECTION

88 · The Last House on the Prairie

1974 Minneapolis Moline G750 · Origin: **Charles City, Iowa** · Company: **White Farm Equipment**

Under the White Farm Equipment umbrella, the Minneapolis-Moline brand had steadily become diluted. The tractors carrying Prairie Gold paint became a mash-up of Oliver and Minneapolis-Moline parts and technology. The last tractor branded Minneapolis-Moline was this G750, and the brand was retired in 1974.

▶ The Minneapolis-Moline G750 was produced from the early 1970s to 1975, and was the same machine as the Oliver 1655. The engine was a Waukesha-Oliver 283-cubic-inch six-cylinder diesel that produced 70 PTO horsepower.
MICHAEL HOOD

89 · A Blue Day for Meadow Green

1976 Oliver 2255 · Origin: **Charles City, Iowa** · Company: **White Farm Equipment**

The market leaders of the early days of tractors suffered mightily under the thumb of White Farm Equipment. The consolidation put Oliver and Minneapolis-Moline in conflict, both struggling to keep their employees and dealers working, and their colors alive. White was also investing more heavily in trucks, and the ag lines were neglected. In the end, ag both brands would disappear. The last Oliver rolled off the line on February 13, 1976.

▶ The last Oliver was a 2255 produced on February 13, 1976. FLOYD COUNTY HISTORICAL SOCIETY

90 · A New Boss

1974 White Field Boss 4-150 · Origin: **Charles City, Iowa** · Company: **White Farm Equipment**

In 1974, White Farm Equipment did a total reboot. In the ensuing year, it ended its three legacy brands of Oliver, Cockshutt, and Minneapolis-Moline—and introduced a new line of machines branded "White." The sleek silver-and-gray line of tractors included several new high-horsepower four-wheel-drives as well as several two-wheel-drive machines that were largely reskinned Minneapolis-Moline and Oliver designs.

▲ The White Field Boss 4-150 was produced from 1974 to 1978. Power was supplied by a Caterpillar 3208.
Marcus Pasveer

91 · Shaking the Ground

Rite Tractor 750 "Earthquake" · Origin: **Great Falls, Montana** · Company: **Rite Tractor**

Dave and John Curtis started a tractor dealership to sell mostly Wagner four-wheel-drive tractors. When Wagner went away, the Curtis brothers had customers requesting something similar, so they began building and selling machines. The machines were component-built, typically using Cummins engines, Spicer drivelines, Clark or Caterpillar axles, and Allison automatic transmissions. Bear in mind each one was custom, so different bits were used on occasion.

The articulation joint was a patented design created by the brothers, and they built their own cabs and transfer case design. As of 2013, they had built a total of thirty-eight machines since starting out in 1973, ranging from 300 horsepower and up. The Rite 750—known as the "Earthquake"—is a 750-horsepower beast that is the flagship of the line.

▼ **Since 1973, a total of three "Earthquake" models have been built by Rite Tractor.**
Peter Simpson
Collection

92 · Triple Power

1975 Steiger Big Jack · Origin: **Fargo, North Dakota** · Company: **Steiger Manufacturing, Inc.**

The CEO of Steiger, Jack Johnson, liked big machines, and much of the company was obsessed with the early 1970s horsepower race. One of its most flamboyant experiments was this triple-engined 750-horsepower machine dubbed "Big Jack." The articulation and traction of the machine was outstanding as all three units could pivot and hug the ground. Engineer Paul Nystuen said that operation and maintenance was horrifically complex, as all three units operated independently. This model never made it past the prototype stage. Steiger also built a twin-engined prototype that was much easier to handle but also was never put into production.

▼ This is Steiger's experimental Big Jack tractor, which was powered by three Caterpillar 3306 250-horsepower engines. Engineer Paul Nystuen, master builder Jerry Joubert, and another man are shown with the machine. Paul Nystuen Image Collection

93 · The Sidehill Stud

1974 J. I. Case 2670 · Origin: **Racine, Wisconsin** · Company: **J. I. Case**

By the mid-1970s, Case had established its niche with high-value, high-horsepower, crab-steering four-wheel-drive machines. The 2470's bigger brother, or should we say its more buff twin brother, was the 2670 Traction King, introduced in 1974. This one is doing what a rigid, four-wheel steer does best, working a sidehill and pulling a straight furrow. The 2470 / 2670 introduced a coordinated crab steer, meaning the operator could control both axles from the steering wheel rather than having to work the wheel and a separate rear wheel control. It was a complex cable and hydraulic setup, but it worked.

▶ **The 2670 would take the Case A504BDT diesel engine to its limit for the day, delivering 189.35 horsepower on the drawbar and 23,618 pounds of drawbar pull on duals.** CASE IH

▶ **This drawing by industrial designer John Mellberg shows off the new look of the White Power line.**
JOHN MELLBERG

94 · The Game Changer

1975 John Deere 8630 · Origin: **Moline, Illinois** · Company: **John Deere**

John Deere was mercilessly on target in the 1970s, and the 8430 and 8630 four-wheel-drives were another direct hit. The big machines offered high-horsepower and comfortable cabs, and allowed John Deere to lead the segment. The engine was a turbocharged and intercooled 619-cubic-inch six-cylinder diesel hooked to a partial powershift transmission.

▶ **In production from 1975 to 1978, the 8630 produced 225.6 PTO horsepower when tested at Nebraska in June 1975.**
GEOFF WING

BIG DEMAND FOR TRACTION AND POWER

The four-wheel-drive market grew rapidly in the early 1970s. Sales of four-wheel-drive tractors above 100 horsepower increased from 28,728 units in 1971 to 62,081 units in 1975.

This chart was part of an International Harvester Company 1976 internal report on the four-wheel-drive market. The report concluded with recommendations that International should develop a 130- and 150-horsepower row-crop four-wheel-drive machine and work with its supplier, Steiger, to develop 350- and 400-horsepower models.

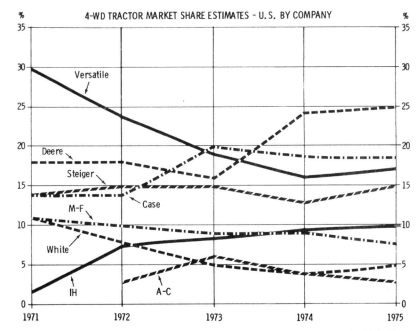

▲ This chart, showing estimated sales, was created by IHC management in 1976. Note that J. I. Case led the industry in 1973, and that the Deere 8430 / 8630 four-wheel-drives made quite an impact in 1975. Also bear in mind most of the IH and Allis-Chalmers machines were contract-built by Steiger. WISCONSIN HISTORICAL SOCIETY

95 · Bicentennial Tractors

Produced: **1976** · Origin: **Varies** · Company: **International Harvester Company, Steiger, J. I. Case**

▲ The J. I. Case 1570 was a powerful, two-wheel-drive machine, built at a time when the company was attacking the high-horsepower market with vigor.
CASE IH

▶ By 1976, the Cub Cadet was a smash hit, cutting grass on suburban and rural lawns across America.
WISCONSIN HISTORICAL SOCIETY 11286

In 1976, a wide variety of makers decided to celebrate America's bicentennial with special paint jobs on their machines. Although each had their own garishly flamboyant flair, Steiger's effort was one of the most entertaining. They hired Spud Sturgis to promote the machine, and he did so with a pizzazz for the dramatic—unmatched then or since.

J. I. Case joined the fray with its high-horsepower, two-wheel-drive Model 1570, and International Harvester produced a Cub Cadet emblazoned with stars and bars. The Steiger Panther II was the beast of the bunch. Massey Ferguson also had a model and wins the prize for most garish— the 1150.

◄ Steiger's edition was as garish as the rest, but with the costume of its promoter, the machine was a full-on 1970s blast to the eyeballs.
SCOTT ANDERSON IMAGE COLLECTION

◄ Steiger's Series III machines were some of the finest four-wheel-drive high-horsepower machines on the market in the era.
PAUL NYSTUEN IMAGE COLLECTION

96 · The Two-Way Tractor

1977 Versatile 150 · Origin: **Winnipeg, Manitoba** · Company: **Versatile**

by Jim Allen

▼ The Versatile 150 debuted in 1977 as the world's first bidirectional (BiDi) tractor. It was articulated, and the driver could operate it facing either direction. This tractor is shown with the controls rotated toward the engine (there really is no "front" or "rear" with a BiDi) and a cultivator is being used as it would be on a conventional tractor. NEW HOLLAND

If you've ever worked a conventional ag tractor, you'll know the business end is at the back. The primary lift and hitch apparatus is there, as are the PTO and the big tires. This traditional layout works fine when the tractor is moving forward, but some implements (most notably snowplows, blowers, certain haying equipment, and swathers / windrowers), as well as some types of mowers, aren't very efficient or don't work that way. This leaves it up to the operator to get neck strain from maneuvering the tractor in reverse.

Even though you can buy loaders and other equipment for it, the front of most tractors is significantly less capable of carrying weight than the rear, and fitting PTOs up front is much more difficult and seldom done. All these drawbacks were much more present in the third quarter of the twentieth century until Dan Pakosh at Versatile Manufacturing in Manitoba, Canada, came up with a solution.

To Pakosh, the answer was simple. Build a tractor that could go forward and back with equal ease, have approximately the same capacity in both directions, and most important, have an operator station that would allow the driver to face either direction. The Versatile 150 debuted in 1977 as a highly maneuverable articulated tractor with a stout hydrostatic drive. It was the world's first production Bidirectional (BiDi) tractor, and it was powered by a four-cylinder Perkins 4.236 NA diesel that made 71 horsepower and 192 pound-foot of torque. The 150 was an immediate hit in grain farming in the northern states and Canada, where wheat was commonly cut and windrowed with a swather to dry before later combining,

rather than being combined off the field. The Versatile 150 was built from 1977 to 1981, and followed in 1982 by the similar Versatile 160. The main difference in the newer model was the use of a 220-cubic-inch Waukesha turbocharged four-cylinder engine that made a few more horsepower than the Perkins.

The Versatile 160 lasted until 1983, when it was replaced by the 256. The 276 appeared in 1985 (painted Ford blue after 1988) and was later revised to become the 9030. The biggest change was in 1998, when a new design based on New Holland's T6070 tractor design was unveiled as the TV6070. Under parent firm CNH Industrial, volumes were deemed too small, and New Holland ceased BiDi production in 2014.

Although it didn't take long for other companies to develop tractors that had similar capabilities to the Versatile / Ford New Holland Bidirectional tractors, the New Holland TV line became the best-known bi-directional tractor on the market.

97 · Looking Forward

1977 Deutz Intrac · Origin: **Cologne, Germany** · Company: **Klöckner-Humboldt-Deutz**

Klöckner-Humboldt-Deutz (KHD) had been in the tractor business for almost sixty years when it launched onto the North American market in 1966. One of its most innovative designs was the forward-control Intrac, which gave a good view of front-mounted equipment and could accommodate a hopper or sprayer on the rear platform.

▼ The forward-control Intrac seated the driver up front for easy operation of front-mounted equipment, such as a mower, and freed up rear load space for a hopper or sprayer.
KHD / SDF

98 · Engine-Forward Design

1977 International TX-178 · Origin: **Chicago, Illinois** · Company: **International Harvester Company**

▼ **This concept drawing of the TX-178 shows off how streamlined the tractor could be with the radiator mounted behind the engine.**
Rich Hale Image Collection

Visibility and cab noise were two major concerns for tractor developers in the 1970s, and this International concept tractor attempted to address both issues by mounting the radiator in the middle of the tractor. This pushed the engine a bit farther forward, which reduced the engine noise in the cab and improved visibility by allowing the hood to be a narrow shark nose. The idea was extensively tested in 1977, and the test engineers found good cooling performance, improved visibility, and better front-end stability. Despite this, the project was scrapped due to significant problems with trash build-up on the air intake as well as driveline vibration. The reverse air-flow concept, however, was not scrapped and would appear on later IH models.

◀ This image from 1977 shows the TX-178 being assembled in the shop. Note the radiator location. RICH HALE IMAGE COLLECTION

◀ This is the finished TX-178 prototype, which was tested extensively in March 1977. RICH HALE IMAGE COLLECTION

99 · Steiger Hits Stride

1977 Steiger Tiger 470 Series III · Origin: **Fargo, North Dakota** · Company: **Steiger Manufacturing, Inc.**

▲ The Steiger Tiger ST450 Series III appeared in 1977, and was revised as the ST470 in 1980. The Tiger was the largest in the lineup, with the Cummins 1150-cubic-inch engine producing more than 450 horsepower. This one is shown in Australia, with a Magnum 7150 lurking in the shadows.

GILL JONES COLLECTION / LEE KLANCHER

Steiger had been cash-positive and growing rapidly throughout the mid-1970s. The four-wheel-drive tractor market was exploding, the management group was sharp, experienced, and aggressive, and sales were Wild West spectacular. The company was able to meet production demands with its new 420,000-square-foot plant in Fargo, North Dakota, the dealer network was stable and solid, and its warranty and service programs were state of the art (Steiger was one of the first companies to offer loaner tractors while your machine was serviced, for example). The sales dollars went heavily back into development, and the Series III machines were refined, comfortable, high-tech, and as rugged as ever.

100 · Canadian High Power

1978 Massey Ferguson 4840 · Origin: **Brantford, Ontario** · Company: **Massey Ferguson**

The Massey contingent's take on the four-wheel-drive market was to go after the same range as Steiger: the 225- to 400-horsepower range of machines. The company introduced a line of all-new machines in 1978, with the 4880 (and later the 4900) at the top of the horsepower heap.

◀ The Massey Ferguson 4840 was produced from 1978 to 1983. The 903-cubic-inch Cummins turbocharged V8 diesel engine produced 210.7 PTO horsepower when tested at Nebraska in 1980. Peter Simpson Collection

101 · A New Line

1978 Allis-Chalmers 8550 · Origin: **West Allis, Wisconsin** · Company: **Allis-Chalmers Company**

Allis-Chalmers released a new, state-of-the-art line of tractors, the 7000 series, in 1974. Using the tooling from that line, it created its own four-wheel-drive to replace the Model 440, which was sourced from Steiger. Introduced in 1976, the 7580 four-wheel-drive was a 186-horsepower entry onto the market. The higher horsepower 8550 came out in 1978, giving Persian Orange a solid lineup in one of the hottest markets in farm history.

▶ Introduced in 1978, the Allis-Chalmers 8550 produced 253.9 PTO horsepower when tested at Nebraska in October 1977. Super T © 2017

102 · Big Roy

1977 Versatile 1080 "Big Roy" · Origin: **Winnipeg, Manitoba** · Company: **Versatile Manufacturing**

▼ **Versatile's Model 1080 is a one-off prototype powered by a Cummins 600-horsepower engine.**
Manitoba Agricultural Museum

The race for horsepower bragging rights continued, and Peter Pakosh and Roy Robinson, the leaders of Versatile, jumped into the fray in 1977. Their engineers put a Cummins KTA-1150 600-horsepower engine in the rear of this four-axle, eight-wheeled behemoth, and dubbed it "Big Roy." The tractor had two articulation joints, and was more than 30 feet long. The vision to the rear was so poor the machine used a closed-circuit television system that displayed the rear view on a 9-inch screen in the dash. The eight tires compacted the ground mercilessly, and, despite a raft of modifications during development, the machine never worked well enough to merit production. The one example built resides in the Manitoba Agricultural Museum.

103 · The Beast

1978 Big Bud 16V-747 · Origin: **Havre, Montana** · Company: **Northern Manufacturing Company**

In 1975, Ron Harmon bought the Northern Manufacturing Company from his neighbor, and quickly grew production and distribution of Big Bud tractors. The company moved into a new facility, and production expanded from roughly twelve to eighty tractors a year. In January 1978, Big Bud released the world's largest tractor: the 16V-747. The custom-built machine was 27 feet long, 20 feet wide, and weighed more than 100,000 pounds when the 850-gallon fuel tank was full.

▼ The Big Bud 16V-747 was powered by a sixteen-cylinder Detroit Diesel engine with more than 760 horsepower. When pulling an 80-foot cultivator, it could work more than one acre per minute.
Michael Hood

104 · The Iron Horses

1978 John Deere 8440 · Origin: **Moline, Illinois** · Company: **John Deere**

By 1978, William A. Hewitt had been at the helm at John Deere for twenty years. Under his more aggressive leadership, the company had spent fifteen years as the world's largest agricultural equipment maker. Its presence had grown globally, and the company was the leader in terms of annual growth. By 1980, 29 percent of the agricultural equipment market would belong to John Deere. The only company to even hold a double-digit share was International, at 14 percent. The rest held 8 percent (Massey Ferguson) or less. The *only* market segment that John Deere struggled with was the under-40-horsepower tractor segment, in which Kubota had more than 30 percent of the market.

The Iron Horse series introduced in 1978 continued to raise standards for the tractor market, with horsepower ranging from 90 to 180, advanced transmissions, and quiet, comfortable cabs.

▼ **The John Deere 8440 was built from 1979 to 1982, was rated for 215 horsepower, and cost $64,000 in 1982.**

Geoff Wing

A SMASHING INTRODUCTION

Chuck Pelly, a well-known industrial designer based in southern California, penned designs for BMW, Porsche, Samsonite, the Disney monorail, and many more. He went to work in the late 1960s for Henry Dreyfuss Associates (HDA) and was one of the last designers to work closely with the old master himself, Henry Dreyfuss.

One of Pelly's projects while at HDA was the design of a new John Deere cab. After months of design work, the first mock-up of the cab was to be presented to John Deere executives on a stage. Pelly recalls taking great pride in the new design, and wanting to present it as it would function: in motion.

"[The mock-up was] made out of paper, and tape, and bent plastic," he said, recalling it was just a visual representation, and not terribly sturdy. "I'm not a good tractor driver but I wanted to present it moving."

When the audience gathered, Pelly intended to drive the machine out for them to see it for the very first time.

"I had all the big John Deere people there down at the bottom of the presentation area. There was a Volkswagen parked there."

Pelly had very little experience driving a tractor, and the machine got away from him and went careening off track, toward the parked car.

"I ran over the front of the Volkswagen," Pelly said. "Of course, it came apart."

He ran it over with one big right tire. The small car came apart like a cheap watch. Pelly also ruptured the fuel tank.

"I drilled a hole in the tank and the battery spun. Everyone hid behind trees."

Pelly figured that was the end of his work with John Deere and HDA for that matter.

"I packed up my stuff and said, 'Oh, okay I know I'm fired.' Instead, the head of John Deere engineering said, 'We have a job for you. We want you to be a rollover test engineer.'"

Pelly laughed recalling the engineer's good sense of humor. The cab would go on as a hit for John Deere, and Pelly would continue to work happily with the company, and would eventually found his own firm doing work for BMW, Steiger, and many others.

◀ **Chuck Pelly captures the chaos of the incident in this hand-drawn illustration.**
Chuck Pelly

105 · Snoopy

1979 International 3388 · Origin: **Chicago, Illinois** · Company: **International Harvester Company**

▼ **The International Harvester 3388 was built from 1979 to 1981. Power was supplied by a 466-cubic-inch turbocharged six-cylinder diesel engine good for 170 PTO horsepower when tested at Nebraska in November 1980.**

Wisconsin Historical Society

Throughout the 1970s, the International Harvester Company struggled mightily with finances and battled through restructuring, among many other issues. The once-dominant industry leader had slid to number two, but had a firm grip on that position. Due to a crippling debt load, the company was fighting for survival. One of the cards it played well during this struggle was the company's incredible sales and marketing prowess. In 1976, company leadership found an opportunity in the 150-horsepower, row-crop four-wheel-drive market segment.

IH former territory manager, Russ Whitacre, recalls a late 1970s meeting where Bud Youle said it was "buy time."

"I took that to mean that they couldn't afford to upgrade the whole line and needed some marketing buzz for a new model using existing components," Whitacre wrote on the Industrial Scenery blog.

With some quick thinking, the engineering group realized they could build such a machine by using the rear end from their new 86 series tractors, commissioning a new front axle, adding an all-new pressure-flow-compensated hydraulic system, and incorporating their engine-forward concept developed on TX-178 to give the machine exceptional weight distribution. The new models—the 3388 and 3588—were dubbed the "2+2" line.

Testing of the new line was done in absolute secrecy. Due to their unique appearance, the machines had to be covered with more than the typical canvas to avoid capturing the attention of the competition. As 2+2 engineer Bill Schubert recalls it, less than a half-dozen prototypes were built and tested all around the United States.

"We actually built a box on a back of a trailer, a semitrailer, that you could expand out with electric jacks, drive the tractor in, close the back up so it was down to the eight-foot width, and the axle bars, both front and rear, which stuck out of holes in the side of the container. That's how we transported it," Schubert said.

The testing went well, and the 2+2s were introduced with great fanfare in early 1979. Led by IH's salesman extraordinaire, Bud Youle, the elaborate introduction at Dick Van Dyke studios in Scottsdale, Arizona, cost $1.2 million in dancing girls, a rendition of a Kenny Rogers song, steak dinners for dealers from around the world, and more. The lavish rollout enticed dealerships to pre-order more than 6,000 machines. Although the initial tractors had a few teething issues, the major problem was how the machine looked and was set up. Farmers were taken aback by the long nose, dubbing the tractor "Snoopy." The 2+2 shined, working fast with relatively light implements. Most people would work them like a Steiger, pulling very large implements slower. Ron Birkey's dealerships in Illinois knew precisely how to set up the machines, and sold more than anyone else in the country. For the rest of the world, however, the 2+2s were a hard sell. Today, the machines are prized for their collectability and utility.

▼ **This pristine example is the first 3588 built.**
Jerry Kuster Collection / Josh Kufahl

TRACTORCADE

The 1970s were a tumultuous decade for the farmer. Market fluctuations, always part of the farming game, were extraordinarily dramatic. Beginning in 1972, the devaluation of the dollar worldwide made grain and beans more affordable overseas, and export demand rose. The harvest was poor worldwide in 1972, which increased demand to historic proportion.

Many farmers met the demand with expensive updated equipment, which fewer farmers working larger plots of land continued. Farm policy didn't help matters. New farm bills passed in 1970, 1973, and 1977 failed to adequately meet the needs of many farmers. The 1977 act, in particular, was perceived as a piece of legislation that would incentivize expensive large-scale production without adequate price support.

The 1977 farm bill prompted a group of farmers in Campo, Colorado, to organize and create the American Agriculture Movement (AAM). According to "The Emergence of the American Agricultural Movement, 1977-1979," by William Browne and John Dinse, on September 6, 1977, the founders of AAM organized a rally in Springfield, Colorado, and called for a strike from farm production. About 140 people attended from neighboring states as well as Kansas and Texas.

Two weeks later, they drew more than 2,000 people to a rally in Pueblo, Texas.

The rallies continued to grow and appear around the country, and they also began to end with a caravan of tractors and trucks parading through the local business district. AAM claimed more than three million farmers participated at the height of their activity.

The group demanded the producers were paid fairly for their crops, and would not produce any crops until their demands were met. The purpose of the movement and the tractorcade was to draw attention to the farmers' plight and force the Congress to act.

The largest tractorcade took place on January 18, 1978, when nearly 3,000 farmers brought a long string of tractors into Washington, DC. The tractors were parked peacefully at a local stadium, and throngs of farmers politely flooded congressional offices. Roughly a third of the crowd stayed in town through the spring. The demonstration prompted Senator Bob Dole to introduce an amendment that passed easily in the Senate. The bill was defeated in the House, and a heavily revised bill passed and was signed into law by President Carter on May 15. Organizers viewed the bill as a sellout.

The cost of the protest for farmers began to become onerous and the movement to strike replaced by the need to plant. In fact, USDA statistics show that production increased that year.

The AAM, frustrated with their lack of results, stepped up their message, and the 1979 tractorcade was a different event. Those that participated were angrier, and 1,500 farmers on roughly 900 tractors participated. The tractors were used to tie up traffic and block intersections. The police eventually confined them to the Mall. A tractor was burned, and the 1979 protest was less well received and more emotionally charged than the 1978 version.

The farm economy took a deep dive in 1979 when the grain embargo depressed farm prices. The next few years would prove to be the most challenging time for farmers in modern history.

◀ In January 1979, farmers who were part of the American Agricultural Movement drove tractors to Washington, DC, in an effort to encourage more equitable farm legislation. DIVISION OF WORK & INDUSTRY, NATIONAL MUSEUM OF AMERICAN HISTORY, SMITHSONIAN INSTITUTION

▲ Tractors occupied the Mall in 1979 for more than two weeks in the dead of winter. Farmers and their tractors stayed in DC after it was over to help repair the damages done to the area. JO FREEMAN, © 1979, WWW.JOFREEMAN.COM

ONLY THE STRONG WILL SURVIVE
1980–1989

The 1980s were the most difficult times for the farm economy
since the Great Depression. Agricultural equipment companies
whose balance sheets were not in perfect order were consolidated,
reduced, or dissolved. LEE KLANCHER

106 · Big and in Demand

1980 Big Bud 525/50 · Origin: **Havre, Montana** · Company: **Northern Manufacturing Company**

Big Bud entered the 1980s with its new Series III tractors, which featured an all-new and much heavier chassis. The bolt-on (rather than welded) engine mounts allowed the user to mount any powerplant from 400 to 800 horsepower, and a Twin Disc powershift transmission kept things smooth. As told in Peter Simpson's book *The Big Bud Tractor Story*, the company's chief engineer, Keith Richardson, learned to repair and design tractors as a farm mechanic working in the field. His prototypes were legendarily rugged and well thought-out, and the firm took great pride in the functionality and practicality of its designs.

Big Bud restructured the company and financing in October 1981, paying back $6 million to its bank and taking out new loans for the future. Sales in July and August 1981 were the best months on record for Big Bud. For the moment, the future was bright.

▼ The Big Bud 525/50 was powered by a Cummins KTA1150. This was the most prolific model in the line, with 150 built.

Jim Allen Collection

107 · From Italy to Kansas

1981 Hesston 980DT · Origin: **Turin, Italy** · Company: **Fiat Trattori S.p.A.**

By Martin Rickatson

In 1977, Italian vehicle giant Fiat acquired Kansas-based forage equipment specialist Hesston. This gave Fiat the opportunity to access the North American tractor market by rebadging the 42–162 PTO horsepower Fiat models with the Hesston name. The center-line front-wheel assist, as seen on the 980DT in the foreground here, had been pioneered by Fiat in 1953.

▶ **Fiat's purchase of Hesston gave it access to the North American tractor market, and its Italian-built machines, bearing the American forage brand, complemented its implement lines.** FIAT

108 · Overkill

1982 Allis-Chalmers 4W-305 · Origin: **West Allis, Wisconsin** · Company: **Allis-Chalmers**

The orange brigade was on the march in the early 1980s. Allis-Chalmers introduced the 60X0 series, a new line of midsize row-crop tractors, and in 1980 and in 1982 followed that up with the 8000 series, a heavily revised rendition of its large tractors. Its flashiest additions were the 4W-220 and 4W-305, two high-horsepower four-wheel-drive machines with twenty-speed transmissions and lavish new cabs. The 4W-220 was discontinued when it was discovered that the 8070 FWA could do just as much work and cost $10,000 less.

▶ **The Allis-Chalmers 4W-305 was introduced in 1982 and produced until 1985. Power was provided by a 731-cubic-inch six-cylinder turbo-diesel.** AGCO

109 · Aussie Innovation

1982 Baldwin DP600 · Origin: **Rooty Hill, Australia** · Company: **E. M. Baldwin**

▼ The Baldwin DP600 is powered by a Cummins 1150-cubic-inch six-cylinder diesel rated for 600 horsepower. The machine was produced only a few years after 1982, and was the most powerful wheel tractor in Australia.

Marcus Pasveer

The world continued to find opportunities for some, including E. M. Baldwin and his sons. The Australian countryside had long been a rich market for high-horsepower machines, particularly four-wheel-drives. Steiger was, and still is, a big name in a land where fields stretch for miles and a single farm can have as much land as a typical American county. E. M. Baldwin & Sons built heavy equipment near Rooty Hill, Australia, and developed a reputation for good-quality sugar cane machinery as well as mining and tunneling vehicles. In 1979, the company turned its attention to large agricultural tractors. Its first machine, the Baldwin Broadacre Tractor, and the model DM 525 won the Australian Design Award in 1982. The company created six different models, and ceased production in the late 1980s.

TECHNOLOGY AND THE FARM CRISIS

The changes taking place on the farm since the tractor was invented came to a head in the early 1980s. When the first farm tractors began to appear in the 1910s, farmers made up 31 percent of the American labor force. They worked 6.4 million farms, averaging 138 acres each.

As the American population exploded, the number of farmers decreased. A wide variety of forces caused this, but the massive improvements in farm technology were critical in this trend, as they allowed a very few to feed many. By 1980, farmers made up 3.4 percent of the American labor force. They worked 2.4 million farms and the average acreage was 426.

The farmers of the 1980s would face the most difficult challenge since the Great Depression. The American economy took a nosedive as interest rates soared. During the 1970s, demand grew and prices fluctuated. With the need to produce critical to their survival, farmers invested in new equipment

▲ **The 1980s were the hardest time on the farm in modern history.** Wisconsin Historical Society 9267

and machines that had become prohibitively more expensive.

Consider the fact that a Wagner four-wheel-drive tractor was considered state of the art in the late 1950s and cost about $20,000. In 1982, the top-of-the-line John Deere 8850 was $120,000. If you adjust those numbers for inflation, the cost of the tractor had doubled.

In the 1980s, only the strongest farmers would survive, which were those who had either large enough operations to buckle down and endure, or such low debt that they could scale back.

Those with debt often had their farms repossessed by the bank.

The agricultural equipment industry suffered along with its customers. Sales were dismal. In 1982, only John Deere showed a positive cash flow on the balance sheet. Every other agricultural company lost money.

These hard times would require the best of the farmers, dealers, and agricultural companies who would make it through.

110 · The Ag Silicon Valley

1982 Panther 1000 · Origin: **Fargo, North Dakota** · Company: **Steiger Manufacturing, Inc.**

One of the smartest things Steiger did was hire keen young people and let them create new technology and ideas. Some of the best and brightest grads from North Dakota State University and elsewhere found a home at Steiger, and quickly fell in love with the creative environment.

Barry Batcheller was one of those young creatives. "It was fun," he said. "Here we are, a bunch of right-out-of-college kids, and Steiger was this incredible place where they would just let you play."

Batcheller came to Steiger after helping start an agricultural communications company, and one of the challenges he faced early in his career at Steiger was designing the control system for a twelve-speed powershift transmission supplied by Twin Disc. That was supposed to become a Steiger standard feature first, but somehow Versatile introduced a model with that twelve-speed powershift.

That move violated an agreement and effectively ended the relationship between Steiger and Twin Disc, and Batcheller then worked with Fuji Techno on the powershift transmission. The engineers at Fuji Techno were very receptive to new ideas, and Batcheller was able to build an entirely new electronic control system for the transmission. That morphed into an electronic control system dubbed the Steiger Electronic Control Center, which was a nerve center for the tractor. "We wrote all the software, we wrote algorithms that did ground speed matching. . . . We did a lot of really interesting things."

At this time, no suppliers existed for such things, and the Steiger team had to hand-build circuit boards, laying out the designs on small wooden pieces, dubbed

▼ **The Puma 1000 was introduced to cope with the difficult market of the 1980s. The machine had simpler features, a smaller engine, and a more modest 190 horsepower.**
MIKE BALLINGER
COLLECTION /
LEE KLANCHER

◄ The Panther 1000 was billed as the world's most advanced farm tractor. The machine backed the claim with an electronic control system, full powershift transmission, sophisticated cab, air ride seat, and six different configurations ranging from 325 to 400 horsepower. Case IH

"puppets," and using black tape. This was the same process used by Steve Jobs and company to build the original Apple computer in a garage in Silicon Valley. Incidentally, at roughly the same time Batcheller was building control boards, the Japanese tractor maker, Kubota, was producing circuit boards and radio-controlled lawnmowers.

The electronic control system debuted on the Panther 1000. Although it had some teething issues, the technology was the first of its kind and a very early forerunner to the electronic controls found on tractors today. Batcheller went on to help found six agricultural technology companies, and became the director of technology at Deere.

Steiger, in the meantime, handled the balance of the 1980s with aplomb. As the market slowed, it increased spending on marketing and public relations, and decreased production. The company also introduced the Puma, a model designed to be a high-value, low-cost machine that would work in the difficult markets of the mid-1980s.

111 · Betting the House

1982 International 50 / 30 88 Series · Origin: **Chicago, Illinois** · Company: **International Harvester Company**

When Brooks McCormick took over the helm at International Harvester Company in the 1970s, hope was high that the founder's descendant would lead the company in the right direction. McCormick made a number of smart moves that helped the company tremendously, including investing in technology, doing some smart reorganization, and cutting back on less-profitable product lines.

▶ **Early design sketch for new IH series.**
GREGG MONTGOMERY
COLLECTION

▼ **The 5088, 5288, and 5488 were the larger models in the line.**
KEN UPDIKE COLLECTION

One of the much-debated moves McCormick made was to hire Archie McCardell away from Xerox to be the new IH CEO (and pay him a record-setting salary). McCardell changed up the management ranks, brought in ruthless cost-cutters, and fought bitterly with organized labor. The latter led to the infamous 1979 strike, which shuttered IH manufacturing from November 1979 to April 1980, and gained not an inch in negotiation.

Although McCardell was clearly terrible dealing with labor, that strike didn't hurt quite as badly as it might have. The farm market tanked about the same time the United Auto Workers went on strike, so the lack of production didn't cost IH as much in sales as one might expect.

He also wasn't able to change the fact that IH's profit margin was dismal. When McCardell started in 1977, John Deere ran at a 7.1 percent margin, Caterpillar was at 7.9 percent, and IH was at 3.4 percent. Those numbers were about the average for each company for the 1970s, and all would drop in 1980. John Deere stayed positive at a 1.1 percent profit, Caterpillar dropped to -2.8 percent, and IH clocked in a horrendous -38.2 percent.

The move McCardell made that paid off—for decades to come—was to invest heavily in new technology. Doing so in tough times required a gambler's swagger, and McCardell had no shortage of that. The red tractor line desperately needed a new transmission, and if the company had any hopes of regaining the number one slot in the market (or, for that matter, holding on to number two), then it needed an exciting new model.

The STS transmission was a new design that would not share any parts with Harvester's previous designs. "When we say 'all new,' we mean we started at the engine flywheel and went from there through the entire powertrain and all of its auxiliary components, including hydraulics," said Glenn Kahle, the engineering vice president at that time, in a 1981 interview published in *Diesel Progress* magazine. IH invested more than $200 million to bring the new transmission to production.

George Vater was the lead engineer on the 50 series transmission. "We started the 50 series tractor right there with a clean sheet of paper and we just started working on everything that we could do to make the tractor more reliable, give us plenty of power to grow," Vater said. "We had many reliability improvements that we made, because we wanted this tractor to last and we wanted to have good reliability."

Although the STS transmission was a huge part of the 50 series tractors and contributed to its excellent fuel efficiency, other notable features included operator comfort, a fuel-saving demand compensating hydraulic system, and an innovative cooling system that helped keep the engine at optimal temperature.

The 50 series tractors were equipped with Melrose Park's proven 400 series engines, and the chassis featured a unique forward air-flow design that pushed cooling air out of the cowling in the front. This new system was pioneered by the Advanced Engineering Group on the TX-178, which had the radiator mounted behind the engine.

The 50 series tractors was introduced in September 1981 as 1982 models, and they did raise the bar for high-horsepower two-wheel-drive tractors. Unfortunately, this was not an easy time to sell tractors, and International was fighting the terrible market as well as massive issues with obtaining financing. The debt load and the market were wreaking havoc with red tractors.

While the 50 series was being introduced, the engineers at Hinsdale were continuing to create and test new tractor designs. In fact, they had the makings of the next machine on the books, even as the time (and money) left for International Harvester was coming to a close.

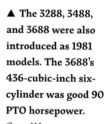

▲ The 3288, 3488, and 3688 were also introduced as 1981 models. The 3688's 436-cubic-inch six-cylinder was good 90 PTO horsepower.
Case IH

112 · Big Green

1982 John Deere 8850 · Origin: **Moline, Illinois** · Company: **John Deere**

▲ **The John Deere 8850 was built from 1982 to 1988, and produced 304 PTO horsepower when tested at Nebraska in May 1982.**

MARCUS PASVEER

Farming's toughest year in the modern era was 1982 (as it was for pretty much everyone in the industry). John Deere had historically weathered hard times with extreme conservatism, but the company in the 1980s had been number one in the United States for nearly two decades. In an uncharacteristically aggressive move, the company unveiled a line of new machines in the fall of 1981. The flagship of the line was the big 8850, with power from a 955-cubic-inch V8 good for 370 horsepower and an advanced quad-range partial powershift transmission. In an equally uncharacteristic John Deere tendency, the big V8 was not ready for prime time and failed all too often, particularly considering the tractor retailed for $120,000. Despite the minor stumble in the line's introduction, the model paid off just fine; John Deere was the only tractor manufacturer in the United States to show a profit in 1982.

THE SPY

When John Deere introduced their new 50 series tractors in the fall of 1981, the red tractor folks were extremely curious to see what the company had in store. Of particular interest was the high-horsepower front-wheel-assist machines, some of which would compete directly with the IH 2+2 line. Bud Youle was one of the key salespeople at that time to drive sales of the 2+2, and he took it upon himself to get an up-close and personal look at the new machines during the dealer introduction at the Superdome in New Orleans.

This curious IH salesman wasn't the first to pull shenanigans to get a sneak peek at the other color's doing. Both sides had personnel out in the field looking specifically for test machines for decades, and stories of engineers tearing apart test machines in the dark and pulling all sorts of devious stunts to get an advance look are fairly common.

So Mr. Bud Youle went down to New Orleans, dressed himself in a John Deere hat and shirt, and headed to the Superdome to see if he could learn a bit about John Deere's new machines.

"I had been in [the Superdome] before," Youle said. "I knew there was a bar up [above the field]. I thought if I could get up to that bar, and I could sit and listen, that's all I needed to do.

"I don't remember exactly how I got in the front door, but I got in. I went to the elevator. I went straight up to the top, to the towers, or whatever they call it. I got out [at] the tower. I walked over, [and] I sat down up there in the dark."

He recalls thinking at the time that if he worked for John Deere and found someone hanging out at the introduction, uninvited, he'd react negatively. His next thought was that he'd better move. As they say in the military, never stay in one spot too long.

"I got myself a Budweiser beer, and got in a different place," Youle explained, "but I could still hear them. Pretty soon, I heard the elevator come up. Two guys got off the elevator."

They were John Deere security people.

Youle knew the gig was up.

"My name is Bud Youle, Harvester Company," he said. "I was down here, and I saw what was going on."

The John Deere folks were not amused. They had spent most of two days trying to find him. Someone had noticed Bud poking around the outside of the dome.

"You're on your way to jail," one of them said.

The men led Youle out.

Youle recalls giving one of the John Deere security guards—who he spent considerable time with in the car when being transported out of town—an earful about the fact that the green side had done the same.

"I'm going to tell you something," Youle said to him. "You remember when we started shipping 2+2s in those boxes? Right above the tractor plant at Rock Island, up on top of that hill was a girls' Catholic school. When the first tractor rolled by that building, there must have been forty-nine cameras taking pictures of that box."

Youle was referring to John Deere employees photographing the sealed boxes used to conceal the test units for the first 2+2s (see tractor 105 for more about that). The point being that employees of all the colors played the game.

"This is an espionage business. I caught you, you caught me," he said.

What he learned was that the new John Deere machines had a larger turning radius than the IH machines. "I always remember, be sure to tell the Deere people that they're going to have trouble wearing out tires, because the tires and the wheels are going to wear off. The guys had a lot of fun with this," Youle said.

The story goes that the IH and John Deere CEOs had a personal talk about the business. Louis Menk was the new IH CEO at the time, and had been brought in to help restructure the company's dire financial straits. You can imagine how pleased the new executive was to deal with a case of employee espionage.

Youle certainly remembers being called into Menk's office. He had an entertaining account of the conversation, as he recalled it forty years after the fact.

"'Mr. Menk, we needed this information.' I said, 'I got our people all geared up and ready to go. We're ready to meet them head-on. We don't have to take a back seat, Mr. Menk.' I said, 'The 5288, the 54 will take care of itself. We can turn shorter.' I said, 'We've got the 72, 74 coming on the 2+2s. We're going to make fools out of them.'"

As Youle recalls, Menk responded genially.

"'Bud,' he said to me, 'this whole thing is handled. Forget it. Let's go on about our business. Let's put them out of business.'

"'Fine, Mr. Menk,' I said. 'That's fine. Go ahead.'"

113 · Rapid Farming

1983 J. I. Case 4890 · Origin: **Racine, Wisconsin** · Company: **J. I. Case**

Under the ownership of Tenneco and the leadership of Jim Ketelsen, J. I. Case was on the move in the early 1980s. Tenneco had vast oil holdings and deep pockets, and at this moment in time, didn't appear to mind investment in its agricultural equipment division.

The company created a joint venture with Cummins to build engines in early 1981. They dubbed the new company Consolidated Diesel Company (CDC), and the engines they built proved to be a strong addition to the company.

In 1982, J. I. Case also merged with David Brown, Ltd., a tractor company that had been purchased by Tenneco in 1972. Case would later abandon that brand, and the Brown family would purchase it back.

One of the innovations Case drove with their new 90 series of tractors was rapid farming—the idea of using less weight on the machine, a smaller implement, and higher speeds. If this sounds familiar, that's precisely how the 2+2 was intended to be used (although only IH dealer Ron Birkey and others appeared to use that to market the machine).

The takeaway here is that J. I. Case had invested a massive amount of money in agricultural equipment line in general, and had just recently invested a large pile of cash into updating and designing its new four-wheel-drive and front wheel assist line. J. I. Case would also lose $68 million in 1983 *alone*. Given all that, can you imagine the company would consider abandoning its shiny new line of machines? Hold that thought.

▼ Sporting a hood scoop, beefed up driveline components, and a 674-cubic-inch Saab-Scania turbo-diesel good for 253 PTO horsepower, the Case 4890 was built from 1979 to 1983. Case IH

114 · New Developments Abroad

1984 Deutz-Fahr DX8.30 Diesel · Origin: **Cologne, Germany** · Company: **Deutz-Fahr**

By Martin Rickatson

Until development of larger Agrotron models in the late 1990s, the largest Deutz tractor was the DX8.30, introduced in 1984. Producing a maximum 217 engine horsepower from its Deutz air-cooled engine, the tractor was a development of the 197-horsepower DX230 launched in 1980, and the 217-horsepower DX250 that replaced it in 1982. All came with standard front-wheel assist.

▶ **Fitted with standard front-wheel assist, the 1984 Deutz-Fahr DX8.30 produced a maximum 217 engine horsepower from its air-cooled engine.** Deutz-Fahr

115 · Versatile Changes Color

1985 Versatile 936 · Origin: **Winnipeg, Manitoba** · Company: **Versatile Manufacturing**

Versatile introduced the Designation 6 series of tractors in 1985, equipped with Cummins engines ranging from 210 to 360 horsepower, with synchro or powershift transmissions. The cab was updated with improved visibility, a new console, and more comfort. John Deere tried to merge with Versatile, but the Department of Justice ruled the merger would give John Deere too much control over the four-wheel-drive market. Ford New Holland purchased Versatile in 1987. The Designation 6 Versatile tractors were painted Ford blue not long after that purchase.

▲ **The Versatile 936 was built from 1985 to 1988, with power from a 855-cubic-inch Cummins and a manual twelve-speed transmission providing 254.4 drawbar horsepower when tested at Nebraska in May 1985.** Library of Congress 092362pu

116 · What Might Have Been

1985 International Harvester 7488 · Origin: **Chicago, Illinois** · Company: **International Harvester Company**

This is the tractor that never really happened. The final development of the "Snoopy" line of articulated IH tractors, the Super 70 (7288 and 7488) offered a refined maneuverable four-wheel-drive tractor capable of working both the open field and between rows of crops. The 2+2s were always a bit of a love-hate relationship for the end user, and they were Harvester's best card in the competitive four-wheel-drive market.

Introduced just as IH's agricultural division was purchased by Tenneco, the plug was pulled on the model and fewer than three dozen of them were built.

The salesman who drove their introduction, International's ever-colorful Bud Youle, recalls the fateful day his phone rang. Robert Greene, the head of agricultural marketing, was on the line.

"All I can remember is when Bob Greene called me up one day. He says, 'You're out of the Snoopy business there,' because they had made the decision. I said, 'Robert, you are making a mistake. You are making a mistake.'

"I learned there's no sense trying to push it. It snowballs up the hill. When there's more up on top of the hill than you, sometimes valor is the better part of discretion, so I backed off."

▼ **In 1984, the 7488 was a high-tech and totally modern machine that served large row-crop and small grain producers well. This is serial number one of only sixteen produced.**
Smith Collection / Lee Klancher

117 · The Last Red Steiger?

1984 International 7388 · Origin: **Fargo, North Dakota** · Company: **Steiger Manufacturing, Inc.**

The International Harvester plan for the high-horsepower market was to work with supplier Steiger to develop several new models. The 7388, 7588, and 7788 were to be those three machines, with the smallest powered by the International Harvester DTI-466 six-cylinder, and the larger two running a bigger V8. IH owned a significant portion of the Steiger company when it was commissioned, but was forced to sell its shares to Deutz AG to raise some much-needed cash. IH's money woes escalated, and the company was forced to cancel its orders for this line of Steiger-built tractors. Only two examples of each were built, and the massive, rare machines are highly sought after collectibles, with the first 7588 built selling at auction for $151,000 in December 2017.

▲ Only two International 7388s were built. This is the second one. The engine is a 798-cubic-inch IH V8 turbo-diesel. Look Collection / Lee Klancher

WHEN THE MIGHTY FALL

The early 1980s were brutal times on the farm and particularly cruel to the longtime industry giant International Harvester. A heavy debt load, poorly managed finances, and a lengthy, ugly labor strike in late 1979 had IH in dire straits. By 1981, despite offering huge incentives—including a promotion that offered a free Scout truck if you bought a new large tractor—the company lost hundreds of millions of dollars. The result was that by 1983, the company was out of cash. It was so bad, some employees were asked to not just take pay cuts, but also to return money.

James Ketelsen was the CEO and chairman of Tenneco, the company that owned J. I. Case. He was a farm boy and liked farm equipment manufacturers. Irv Aal, a fellow farm executive and also an IH executive, called Ketelsen and suggested they discuss an acquisition. Ketelsen agreed, and met with just a few key executives from Harvester for only a

couple of hours. They came out of the room with an agreement.

On November 26, 1984, the deal was announced to the press. The International Harvester agricultural division would be purchased by Tenneco, and the company merged with J. I. Case to create a new agricultural company, Case IH.

▲ This early attempt at a Case IH logo simply put the two old logos together. It was only used for a short time. Case IH

118 · The Italian Contingent Expands

1985 Hesston 180-90 · Origin: **Turin, Italy** · Company: **Fiat Trattori S.p.A.**

Hesston's largest model, the 1880 with 162 PTO horsepower, was superseded in 1985 by the restyled 180-90. Although outward appearance—save for a little restyling—was similar, underneath the 180-90 and its smaller siblings gained all-new engines, transmissions, and rear ends. Two years after launch, a four-step powershift option was added.

▶ Like its 90 series siblings, the Hesston 180-90 was outwardly similar to the model it replaced. The tractor, along with other 90 models, received a new engine, transmission, and rear end.

HESSTON / FIAT

119 · Orange Mix

1985 Allis-Chalmers 8070 · Origin: **West Allis, Wisconsin** · Company: **Allis-Chalmers**

Following its purchase of Allis-Chalmers's ag equipment business in May 1985, German firm Klöckner-Humboldt-Deutz (KHD) subtly relabeled 107–170 horsepower Allis-Chalmers 8010, 8030, 8050, and 8070 tractors with "Deutz-Allis" decals. They were otherwise unchanged from those introduced by Allis-Chalmers in 1982, and remained in production until the West Allis factory's closure in December that year.

▶ Between May 1985, when it purchased Allis-Chalmers's ag business, and December of that year when the West Allis tractor plant was closed, Klöckner-Humboldt-Deutz (KHD) continued producing the 107–170 horsepower Allis-Chalmers 8010, 8030, 8050, and 8070 models with new "Deutz-Allis" decals.

JIM ALLEN COLLECTION

120 · Blue Blends

1986 Ford 7810 · Origin: **Basildon, England** · Company: **Ford New Holland**

Ford acquired Sperry New Holland in 1986, and it was one of the easier blends of colors of the era. New Holland had a long history of building high-quality hay and forage implements as well as combines, and Ford had an up-to-date line of tractors. Ford New Holland picked up Versatile a year later, and the company that resulted had a strong line of equipment. The Ford midsized tractors continued to be built at the company's plant in Basildon, England.

◀ **The front-wheel-assist 7810 and its companion, the 8210, were released in 1986. Both the models were branded Ford, although the 7810 later morphed into the Mexico-built 7810S.** New Holland

121 · Green No More

1986 Steiger · Origin: **Fargo, North Dakota** · Company: **Case IH**

In 1986, Steiger was doing as well as anyone in the industry, at least according the people at the helm at the time (and the financial reports support their claims). The banks, however, saw things differently, and abruptly pulled the company's financing. Steiger had no choice but to sell, and CEO Irv Aal called Jim Ketelsen, the head of Tenneco, who quickly agreed to purchase the Steiger business. The Department of Justice believed the merger gave Case IH too much four-wheel-drive market share and, invoking anti-trust legislation, blocked the sale. Aal flew to Washington DC to argue that the merger was the only way for Steiger to survive, and that several thousand jobs in Fargo were at stake. He was right, the DOJ relented, and Steiger would live on to build more high-quality four-wheel-drive machines.

◀ **Irv Aal (left) with the Puma 1000, one of the last new models created by Steiger before it was purchased by Tenneco in 1986. Aal was the CEO of Steiger at that time, and played a pivotal role in keeping the brand alive.** Case IH

122 · Tracking Innovation

1987 Caterpillar Challenger 65 · Origin: **Peoria, Illinois** · Company: **Caterpillar, Inc.**

By Scott Garvey, Machinery Editor, excerpted from *Grainews*

In the late 1970s, Caterpillar had been selling a limited number of Special Application (SA) crawlers to farmers. The high-horsepower, articulated, four-wheel drive tractors were eating into demand for those crawlers, so Caterpillar executives decided to develop their own four-wheel-drive agricultural machine.

While research and development took place on the Cat wheeled ag tractor program, Dave Janzen, an engineer working at the company's Peoria Proving Ground in Illinois, was redesigning the SA crawlers to make them more appealing to farmers. At the proving ground he modified a D4SA specifically for the farm, and greatly improved its field performance.

The D4SA still suffered from the limitations of steel tracks and having the same "jerk" steering system common to most crawlers of the day. Those were big drawbacks in an agricultural environment.

As Janzen continued working on his D4SA, Ron Satzler, another Cat engineer, was experimenting with the removable, cable-reinforced rubber tread developed for beadless tires by the company's rubber products division. The tread was about the same size as a steel crawler track. The rubber belts did not get approval for use on the construction equipment, but the team came to realize that Janzen's D4SA ag tractor provided an ideal use for it.

"It was obvious to everyone that the rubber track should be installed on the D4SA," Janzen said. The team dubbed tractors using the new rubber belts, Belted Ag Tractors (BAT). Janzen's original D4SA was fitted with rubber treads and taken to a local farm for field trials. The farm owner used Versatile tractors. The belted prototype was painted the same rust red as a Versatile four-wheel-drive tractor, in the hope that anyone who saw it working in the distance would mistake it for one of the farmer's Versatiles. That red color earned the prototype the name "Red D4 BAT."

Satzler and Janzen were now working together full-time on the belted ag tractor's development.

Red D4 BAT was fitted with a thirteen-speed truck transmission, a track suspension, and a differential steering system. Satzler's design remains the basis for steering systems on the Challengers to this day.

After being fitted with all the design updates, the Red D4 BAT showed promise. Cat found itself in

◄ **This cardboard model shows the initial design objective for the Caterpillar wheeled tractor project early on in the development phase.** AGCO

a difficult financial position, however, and research and development money was tight. Executives had to choose between the wheeled tractor and belted tractor production.

Even though the wheeled tractor was virtually ready for production, executives canceled it in favor of the belted tractor after evaluating its potential, much to the shock of many engineers who had been working on the wheeled tractor.

The body style on the new Challenger 65 had been developed for the wheeled tractor by an industrial designer who had previously worked for General Motors.

In 2001, Caterpillar management decided ag tractors were no longer a good fit for the company. The Challenger tractor brand was sold to AGCO. The original Red D4 BAT and the remaining wheeled prototype tractor were donated to Iowa State University when ownership of the brand changed.

◄ **A cutaway view of the components in the original Challenger 65 design.** AGCO

123 · Lock and Load

1988 Case IH Magnum 7140 · Origin: **Racine, Wisconsin** · Company: **Case IH**

The Case IH merger was one of the most difficult transitions for any agricultural company of the 1980s. J. I. Case and International Harvester were not terribly complementary lines of machinery, and the two groups had different ethics and opinions about their position in the market.

"There was always, for several years, what we called 'Us' and 'Them,'" IH engineer Ben McCash said in 2013. "For some people it never went away."

The dealership network also had lots of overlap, and many dealerships would be forced to close their doors. The fallout for the way this merger was handled, and the inevitable conflict that occurred, lasted for decades.

The way to heal that wound and move forward was with a new tractor. Thankfully for all involved, the money former IH CEO Archie McCardell lavished on research and development in the last days of IH paid dividends.

Shortly after the merger, Tenneco leader Jim Ketelsen met with his new chief engineer, Dr. Glenn Kahle, to discuss what he could do going forward. Kahle told him they had a new 100-plus-horsepower tractor on the drawing board. They had a new powershift transmission and some other bits good to go, not to mention lots of ideas—including a brand-new cab design—that Harvester hadn't been able to fund.

"How soon can we get it out?" Ketelsen asked. Kahle said two years, and as best he could determine, about $40 million was required to finish the work on the new model. Ketelsen quickly agreed, giving Kahle a fistful of dollars and a mandate: transform the final Harvester tractor into the first new tractor produced by Case IH.

▼ The original Magnum appeared in 1987. This example was sold at auction in spring 2018, billed as the first retailed Magnum 7140, with a letter of authenticity from Case IH. Mecum Auctions / Lee Klancher

Those funds were fuel for Kahle and his engineers to excel. One of the key things that made the Magnum was this group's level of dedication to making sure the tractor was done right. In December 1986, the Japanese gear supplier, Okubo Gear Co., Ltd., was having issues interpreting the drawings and manufacturing questions. Wanting to ensure the company knew the expectations for delivering the finished parts, Kahle sent over one of his lead engineers, George Vater, to fix the issue. The complication? This was right before Christmas.

"I think we were going on a Christmas holiday and they told me, 'They want you over in Japan tomorrow.' So I had to go down to the Japanese consulate, get my visa and go to Japan the next day," Vater said. "I was there from the 21st of December to I think about the very last day in December."

This was just one example of the dedication it took to compete—and survive—in that era.

For the employees who made a living wage making the machines, the dealers who survived by selling and servicing them, and the farmers who fed their families using the machines, times remained very hard. Fresh new blood in the form of a class-leading machine was necessary for any of these groups to maintain their way of life.

The Magnum would prove to be exactly what was required. With a great new look, innovative internals, and well-tested components, the Magnum was a resounding success. The tractor's CDC engine produced outstanding fuel economy and horsepower. The look was strong and bold, and the tractor raised the bar for high-horsepower two-wheel-drive machines. The cab also set new standards—and that was an area red tractors had been behind on from the 1960s.

Despite its success, the times were rough enough that the Magnum alone couldn't save Case IH. The new model was introduced in August 1987—near the very bottom of the second-worst time in recorded history for the farmer. Throughout 1988, pundits and industry experts forecast resurgence and growth for the farm economy. Droughts and other forces kept that from happening.

The company's survival would be dependent on a farm resurgence. But Case IH had loaded its larders with an agricultural weapon that was good enough to overcome the doubts of those who saw the world through paint colors.

"The success of that tractor saved the company, and it not only saved it but it solidified us as a true company, it brought the two sides together," Steve said. "Both the Case dealer and the IH dealer, or the Case employee and the IH employee, could feel just as proud about that product as the other. And it really solidified our company."

▼ The new Magnum's sleek look takes center stage in this concept drawing. CASE IH

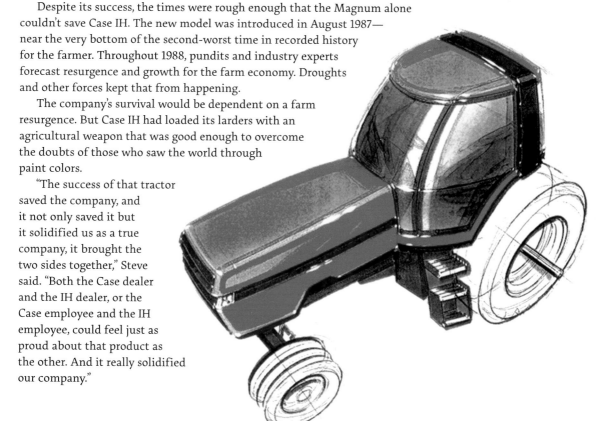

124 · Big Power

1988 White Field Boss 4-270 · Origin: **Coldwater, Ohio** · Company: **White–New Idea**

▲ The White 4-270 was powered by a Caterpillar 3306 638-cubic-inch turbocharged and intercooled diesel engine good for 239.3 PTO horsepower when tested at Nebraska in April 1985.

Marcus Pasveer

White Farm Equipment's 1980s history is more complicated than the plotline of *Twin Peaks*. Money ran out in 1982, and it had to stop production for much of the year, but scraped up some cash and built a few tractors starting again in 1983 and on. White Farm Equipment (WFE) faced a forced bankruptcy in 1985, and was acquired by Allied Farm Products later that year. Allied owned New Idea, and combined the two to form White–New Idea. Stay tuned, there's more. (No dreams, dwarves, or one-armed bandits, though. Sorry.)

125 · The Sales Machine

1988 John Deere 4955 · Origin: **Moline, Illinois** · Company: **John Deere**

In 1988, John Deere released the 4055 series of tractors, ranging from 117 to 222 horsepower. The revised machines boosted horsepower, with updated drivelines to match. Although this was refinement more than revolution, the new machines were popular. The big news for the year? John Deere topped the sales record set in 1979 with the new line.

▲ **The new-for-1988 4955 was powered by the 466-cubic-inch six-cylinder turbo-diesel, good for 210 PTO horsepower when tested at Nebraska in 1989.** JOHN DEERE ARCHIVES

126 · The Green Line

1989 Deutz-Allis 9170 · Origin: **Cologne, Germany** · Company: **Klöckner-Humboldt-Deutz**

▶ The Deutz-Allis 9150/70/90 tractors, which were built in White's Coldwater, Ohio, plant, featured a five-post cab that eliminated the front corner pillars to improve forward vision.
DEUTZ-ALLIS

With the Allis-Chalmers-based 8000 series phased out following the West Allis plant's closure, KHD needed new range-toppers to supplement its smaller models sourced from Deutz-Fahr's German factory. Built in White–New Idea's Coldwater, Ohio plant, the 9150/70/90 featured a five-post cab for improved visibility, with no corner pillars, and the single door closing in line with the exhaust.

127 · Smooth Operator

1989 Renault 155.54 TZ · Origin: **Le Mans, France** · Company: **Renault Agriculture**

The apparent need for a fixed rear end to ensure implements—especially hitch-mounted ones—maintained a consistent depth when in work, meant that for many years tractor operators had to endure a rough ride on top of an unsuspended axle. With its Hydrostable system fitted to TZ models, French manufacturer Renault—which sold its tractor business to Claas in 2003—was the first mainstream maker to introduce cab suspension.

▶ Renault was a pioneer in improving operator comfort by cushioning not just the tractor seat, but the entire cab. Claas inherited its suspension technology when it bought the firm's ag operations.
RENAULT AGRICULTURE

128 · The All-American Machine

1989 White American 60 · Origin: **Coldwater, Ohio** · Company: **White–New Idea**

By Jim Allen

Until 1989, White had a deal with the respected Japanese tractor manufacturer Iseki to build its smaller units in the 30 to 75 horsepower range. The Isekis were good tractors and nobody said differently, but White wanted an American-made smaller tractor. The eventual result was the Model 60 and Model 80 tractors, dubbed the White American Series.

Available with optional front-wheel assist, the models 60 and 80 had CDC engines that produced more than their factory ratings of 60 and 80 horsepower. Many owners and technicians have said both those ratings were wildly understated and that 90 horsepower from the White 80 was fairly common. Retired engineers from White acknowledge this with a wink and a smile.

The normal color scheme for the 60 and 80 was silver, but special editions in Oliver Green, Cockshutt Red, and Minneapolis-Moline Yellow were offered. Unconfirmed production numbers were listed as follows: 162 silver 60s, 435 silver 80s, 91 green 60s, 165 green 80s, 19 red 60s, 19 red 80s, 8 yellow 60s, 9 yellow 80s.

The American 60 and 80 models were fine tractors that were lost in the shuffle when the newly formed AGCO acquired the White tractor lines in 1991, later getting the rest of White Farm Equipment. Once AGCO was in control, the 60 and 80 were discontinued and replaced for 1992 by tractors built by European SAME-Lamborghini-Hürlimann.

▼ The White American 80s were available in four color schemes, each invoking a heritage White brand. The tractors were made entirely in America.
JIM ALLEN COLLECTION

129 · Case IH Maxxum

1992 Maxxum 5150 · Origin: **Neuss, Germany** · Company: **Case IH**

Introduced in 1989 as 1990 models, the Maxxum brought the reliability and power of the Magnum to midsize tractors with 77 to 94 horsepower. Positioned as a direct competitor for the highly respected Ford midsize tractors, the machines were versatile, powerful, and well received. The Maxxum's CDC engines, sixteen-speed semi-powershift transmission, high-capacity lift, and updated cab were strong points. The new models were also another product of the research and development done during the last days of International Harvester—the European engineering team had a midsize machine well along before the merger with J. I. Case in 1984.

▶ **The Maxxum 5150 brought a bit more horsepower to the line in 1992. Rated for 132 PTO horsepower, it was produced from 1992 to 1997. This Maxxtrac edition was sold in Germany.** CASE IH

▲ This full-size example shows off the high-visibility cab developed at International.

THE REBOUND
1990–1999

In the 1990s, the farm economy (finally) came back to life after a long, difficult decade. Farmers were seeing increased revenue and began replacing aging machines (like this 1970s Ford) with much-improved new machines. MARCUS PASVEER

130 · Fast and Furious

1990 JCB Fastrac 145 Turbo · Origin: **Rocester, Staffordshire, England** · Company: **J. C. Bamford Excavators, Ltd.**

▶ **The JCB Fastrac was capable of going up to 40 miles per hour, and was a hit when introduced in 1991. The tractor was later featured as Jeremy Clarkson's favorite on the popular show *Top Gear*.**
JCB

Best known for backhoe loaders and telescopic handlers, JCB previewed its High Mobility Vehicle in 1990. A year later, the new model was launched as the 117-horsepower Fastrac 125 and the 140-horsepower Fastrac 145. Designed with full suspension and a top speed of 40 miles per hour, yet still capable of plowing and performing other tractor tasks, the range has since broadened to encompass models from 160 to 348 engine horsepower, all with CVT.

131 · The Glass Wizards

1995 Zetor 8540 · Origin: **Brno, Czech Republic** · Company: **Zetor a.s.**

▶ **The Zetor 8540 was produced from 1995 to 1999. The Zetor 254-cubic-inch four-cylinder diesel engine was rated for 80 horsepower.**
ZETOR

Czech manufacturer Zetor had long been known for the classic wide-cabbed design of its tractors—offering the possibility to carry one or two passengers—when in 1991 it completely revamped the look of its midrange machines. Featuring sleek new cab and hood designs, the 75-105 engine horsepower 7520/40 to 10520/40 were also marketed in certain southern hemisphere markets under the John Deere name and livery.

132 · The Tool Carrier

1991 Fendt Xylon · Origin: **Marktoberdorf, Germany** · Company: **Fendt**

For many years, German tractor maker Fendt produced, alongside its main tractor range, a line of "toolcarriers," which featured the four-cylinder engine mounted under the cab, clearing the front end for mounting inter-row hoes, spray tanks, and other tools. With the 1991 launch of the 140 engine horsepower Xylon, it took this a stage further, moving the cab to a center point and fitting equal-sized wheels.

▶ **Fendt toolcarriers featured a four-cylinder engine beneath the cab. The 1991 Xylon developed this concept by moving the operator's accommodation forward to a central position between the tractor's equal-sized wheels.**
Fendt / AGCO

133 · The Reboot

1992 Magnum 7150 · Origin: **Racine, Wisconsin** · Company: **Case IH**

In the early 1990s, Case IH decided to create the first 200-horsepower two-wheel-drive row-crop tractor. The company tweaked and boosted, and then put together a blind test at its proving grounds near Casa Grande, New Mexico. The first rendition received a decidedly blah response, so the Case IH engineering team spent another year in development and repeated the blind test. The more muscular version earned a big thumbs-up, and the 7150 hit the market rated for 215 PTO horsepower, but producing a rock star 240 PTO horsepower when tested at Nebraska in November 1992.

▶ **The Magnum 7150 was produced in limited numbers in 1991, and was in full production in 1992.**
Ken Updike Image Collection

134 · Genesis

1993 Ford 8970 · Origin: **Detroit, Michigan** · Company: **Ford New Holland**

Ford introduced an all-new line in 1993, featuring a brand-new engine, an innovative steering system, and European styling. Front-wheel assist (FWA) was one of the hottest options in the 1990s, and steering the machines was a problem for everyone. The 70 series introduced SuperSteer, which pivoted the entire front bolster as well as the wheels, and set a new standard for FWA turning radius. The new 70 series was well received.

▶ **The 8970 was introduced for 1993 and produced until 2002. The largest of the line, the 456-cubic-inch turbo-diesel put out 224.6 PTO horsepower when tested at Nebraska in May 1996.** New Holland

135 · Brand Preservation

1994 Massey Ferguson 375 · Origin: **Duluth, Georgia** · Company: **AGCO**

▶ **The Massey Ferguson 375 was in production from 1987 to 1997, and was built in Coventry, England.** AGCO

The Massey Ferguson brand has a deep history that, interestingly, was preserved by acquisition. In 1990, AGCO was formed in a management buyout of Deutz-Allis from KHD. AGCO added Hesston and White in 1991. In 1994, AGCO bought Massey Ferguson as well as McConnell Tractors. One of the interesting trends that began in the mid-1990s was to preserve brands rather than try to merge them (the J. I. Case and International merger being the prime example of a difficult merger). The result? Massey Ferguson lives on.

136 · Visibly Different

1995 John Deere 8400 · Origin: **Moline, Illinois** · Company: **John Deere**

John Deere's new 8000 series signaled comfort and convenience as the hot button issues of the day when it gave the line's new cab and control system top billing. The four new models were the 8100, 8200, 8300, and 8400.

▼ **Introduced in 1995, the John Deere 8400 featured 224 PTO horsepower and was built until 1999.**

JOHN DEERE ARCHIVES

137 · New Divisions

1995 Caterpillar Challenger 35 and 45 · Origin: **Peoria, Illinois** · Company: **Caterpillar, Inc.**

The Challengers proved that tracks worked, and Caterpillar was so bought into the concept that they started a new agricultural division in February 1996. The row-crop Challenger 35 and 45 appeared in 1995, with new styling and big Caterpillar turbo-diesel engines. The larger 225-horsepower Caterpillar 55 followed in 1996. All three were produced until 1998. In 2002, Caterpillar sold off the Challenger agricultural track division to AGCO.

138 · Quadrophonic

1997 Case IH Steiger Quadtrac · Origin: **Fargo, North Dakota** · Company: **Case IH**

▼ **This 1992 concept drawing shows the early track system grafted to a Steiger four-wheel-drive.** BRYAN GARBERG COLLECTION

The Cat Challenger went directly after the high-horsepower four-wheel-drive market. In places like the Red River Valley with sticky soil and big acreage, tracks were the bomb. Led by a group of former Steiger engineers based in Fargo, Case IH approached the puzzle with four tracks rather than two. With a microscopic research and development budget, the group put together a prototype and tested it in the dark in a hidden river valley only a few miles south of the Canadian border.

The infamous Robert Carlson, known industry-wide as "Tractor Bob," was CEO of Case IH while this was going on. The former John Deere executive knew a winner when he saw one, and pushed the engineering group to roll out a prototype at the 1992 Farm Progress Days. The machine was a sensation, but development of the track system took five long years before seeing production in 1997. The Quadtrac was a hit then, with an improved ride, less ground compaction, and easier transport (as the tracks were narrower than dual or triple tires). Today most of the high-horsepower tractors sold are on tracks rather than tires.

◀ The early Quadtracs bore no model designation—later ones had 9370 or 9380 on the hood. Steiger branding was appropriate, as engineering was done by the four-wheel-drive team in Fargo.
McCann Collection / Lee Klancher

139 · Track Wars

John Deere 9620T · Origin: **Moline, Illinois** · Company: **John Deere**

John Deere entered the tracked market in 1997 with the 8000T models, and was met with furious resistance. Caterpillar quickly filed a lawsuit against Deere, claiming the new machine infringed on its patented rubber-track system. That suit was dismissed on September 1, 1999. John Deere continued production of its dual-track machines into the late 2000s.

◀ John Deere entered the high-horsepower tracked agriculture machine market in 1997. This is the 9620T, a later rendition of the line.
Jim Allen Collection

140 · CVT Pioneers

1995 Fendt 926 Vario · Origin: **Marktoberdorf, Germany** · Company: **AGCO**

▶ **Fendt was the pioneer behind successful application of CVT technology to tractors, with the 1995 launch of the hydromechanical Vario continuously variable transmission on its 926 Vario.**
Fendt / AGCO

The introduction of semi- and full-powershifts had made tractor operation easier, more pleasant, and more efficient. Along with this, hydrostatic transmissions had found a niche. In 1995, Fendt introduced a hydromechanical, continuously variable transmission (CVT) on its 260 engine horsepower 926 Vario. This introduction marked the start of widespread industry adoption of CVT technology. AGCO would purchase Fendt in 1997.

141 · Dressed to Kill

1995 Deutz-Fahr Agrotron · Origin: **Cologne, Germany** · Company: **SAME Deutz-Fahr**

After spending big to get deeper into North America with its 1985 purchase of Allis-Chalmers's ag business, before exiting just five years later by selling to the management buyout that formed AGCO, KHD's next move was to completely redesign its tractors. This move produced arguably the most futuristic look on the market. Launched in 1996, the bubble-cabbed Agrotron line initially spanned 74-256 engine horsepower. Italy's SAME acquired Deutz-Fahr in 1995.

▶ **After exiting North America by selling its Deutz-Allis business to the founders of AGCO, KHD completely re-engineered its German-built tractor offering to the world, creating the bubble-cabbed Agrotron design.**
SAME Deutz-Fahr

142 · Four-Wheel-Steering Magic

1997 Claas Xerion 2500 · Origin: **Harsewinkel, Germany** · Company: **Claas**

By Martin Rickatson

Introduced in 1997, the Claas Xerion four-wheel-steer CVT tractor was initially designed both to take on traditional tractor roles and to accommodate wraparound grain / beet / forage harvesting units. Issues with output, cost, and body-swapping speed refocused subsequent development on providing an alternative to articulated-steer high-horsepower tractors. A hydraulically reversible cab option allowed operation facing rearward to work with, for example, triple mowers.

▶ **The Claas Xerion four-wheel-steer CVT tractor was initially designed both to take on traditional tractor roles and to accommodate wraparound harvesting units, but subsequent development focused on providing an alternative to articulated-steer high-horsepower tractors.** CLAAS

143 · Reversible Upgrades

1998 New Holland TV140 · Origin: **Winnipeg, Manitoba** · Company: **CNH Global**

By Jim Allen

For 1998, the New Holland TV140 tractors appeared with the bi-directional tractor's biggest makeover since it debuted in the 1970s. The unit was enlarged overall and hitch, lift, and PTO apparatus were fitted optionally to both ends. You could run implements not only at either end, but also simultaneously. As before, the entire control station rotated 180 degrees to face fore or aft as needed but the controls and comfort features were light-years ahead of the early tractors.

When the Versatile plant had to be sold in the late 1990s to avoid anti-trust problems, New Holland sold it to Buhler Industries in 2000, and BiDi production was moved down to Fargo, North Dakota, where it continued until 2014. Buhler revived the Versatile name but the BiDi stayed with New Holland. The TV145 debuted in 2004 as an evolved version of the TV140 and continued in production into 2008.

▶ **The power on the new TV140 was upgraded significantly with the addition of a 456-cubic-inch, 107 PTO horsepower New Holland "Genesis" six-cylinder turbo diesel.** NEW HOLLAND

THE TWENTY-FIRST CENTURY
2000–Present

Modern tractors are massive machines that get work done at an unprecedented rate. They also feature advanced electronics that control tractor systems and manage planting and harvesting efficiency.
Lee Klancher

144 · Refreshing Lines

2000 Case IH Steiger STX 450 · Origin: **Fargo, North Dakota** · Company: **Case IH**

By the mid-1990s, the Steiger line of four-wheel-drives was getting dated. The Steiger name had been back on the machine since 1995, and four-wheel-drive engineering was still done in Fargo. Case IH worked its magic with the new-for-2000 STX, which featured new styling, hydraulics, steering system, and transmission. "We basically designed an entirely new tractor," said Case IH product manager Greg Thode.

▶ **The Steiger STX series was introduced in 2000, with the STX succeeding the original STX440 in 2002.**
Case IH

145 · Good Eye

2002 New Holland TG285 · Origin: **Racine, Wisconsin** · Company: **CNH Global**

New Holland made a mark with its distinctively designed TG line of tractors introduced in 2002. Styled by industrial designer Gregg Montgomery, the new CatEye front end drew attention and the solid CDC engines took care of business. In 1999, Case IH and New Holland were merged to form CNH Global NV. The two brands initially shared some common base platforms for their product lines, but remained independently engineered, marketed, serviced, and sold.

▶ **The New Holland TG285 was built from 2002 to 2006, with power from a CDC 8.3-liter diesel engine.**
New Holland

146 · Leader of the Pack

2004 Fendt 930 Vario TMS · Origin: **Marktoberdorf, Germany** · Company: **AGCO**

When Fendt entered the CVT tractor market in 1995, it was with a single 260-horsepower flagship. It wasn't long, though, before the German maker expanded its line of CVT models, and today it offers only the one transmission type.

▶ **Fendt's 300 engine horsepower 930 Vario TMS was the successor to its original Vario CVT tractor, the 926, as the flagship of its range, taking the transmission into new power territory.** AGCO / FENDT

147 · Cross-Pollination

2005 Massey Ferguson 8480 Dyna-VT · Origin: **Beauvais, France** · Company: **AGCO**

With the 1990 formation of AGCO, and the firm's subsequent acquisition of Massey Ferguson (MF) in 1995 and Fendt in 1997, cross-fertilization of technologies between the formerly independent firms became possible. One result was the adoption by MF of the continuously variable transmission technology developed by Fendt. Labeled Dyna-VT, it was made available on tractors right up to this flagship 240 PTO horsepower 8480.

▶ **Massey Ferguson's 8480 Dyna-VT tractor benefited from cross-fertilization of technology following parent company AGCO's acquisition of German firm Fendt, gaining a version of the latter's Vario CVT.** AGCO

148 · Small and Powerful

2005 Kubota M125X · Origin: **Osaka, Japan** · Company: **Kubota Corporation**

Kubota had been producing small, smartly engineered tractors for the American market since the 1960s. By 1981, it owned the under-40-horsepower domestic tractor market, with more than a 30 percent market share. In 2007, this was Kubota's largest tractor, the M125X. The five-cylinder engine cranked out a nice 256 pound-feet at 1200 rpm, with a 2400-rpm redline. Backing that up was a sixteen-speed gearbox

▶ **The Kubota M125X's 356-cubic-inch five-cylinder diesel was direct injected with a Denso inline mechanical injection pump and a wastegated turbo, cranking out 125 gross flywheel horsepower (114 net and 103 PTO).**

JIM ALLEN COLLECTION

with "Intelli-Shift," which had three selectable modes: road, field and shuttle shift. Optional was a twenty-four-speed creeper gear transmission for loader operation. It had standard FWA (which Kubota called "four-wheel drive") featuring a front axle that used bevel gears at the knuckle rather than CV joints. The tractor was commonly sold with a well-appointed cab, but an open-station version was available. Kubota offered a wide variety of implements for the tractor as well.

149 · The Wheel Train

2007 Fendt TRISIX · Origin: **Marktoberdorf, Germany** · Company: **AGCO**

In 2007, Fendt, a German division of AGCO, debuted a concept tractor called the TRISIX Vario. The six-wheel-drive system added traction and pulling power without resorting to tracks or duals, both of which were not well suited to European farming. The TRISIX could mount implements front and rear, and the big, closely spaced radials offered very low ground pressure. The six-by-six configuration put the big Fendt somewhere between an equivalently powered tracked unit and a dual-wheel articulated tractor, but with more road-going capability than either. The TRISIX couldn't match the turning circle of the articulated or the tracked tractor, but the sidehill performance was better than a tracked tractor. The experiment garnered a lot of attention for Fendt, and perhaps a few steps on the ladder of knowledge, but didn't result in a production unit.

▶ **The TRISIX Vario was a one-off six-wheel drive prototype built by Fendt in 2007. Power came from a MAN 12.4-liter inline six that cranked 540 maximum horsepower and a bit over 1800 pound-feet of torque.**

JIM ALLEN COLLECTION

150 · Breaking Big

2007 Challenger MT975B · Origin: **Jackson, Minnesota** · Company: **AGCO**

With 570 horsepower, the Challenger MT975B was the most powerful wheeled tractor you could buy in the early part of 2007. It had a tracked brother, the MT875B, which was the first tractor to break the 500 drawbar horsepower barrier. In 2002, Caterpillar had sold its Challenger track tractor line to AGCO, which went on to develop wheeled Challenger models. The MT975B was powered by a Cat C-18, 24-valve, and 1,106-cubic-inch six-cylinder torque monster that was the largest tractor engine offered at the time. It was rated at 570 horsepower and 1469 pound-foot, but the MEUI (Mechanically actuated Electronically controlled Unit Injector, pronounced "Meuey") injection system would automatically boost power up to 615 horsepower and 2040 pound-feet for short periods under the right conditions. For that reason, it's often called a "600 horsepower" tractor. The engine was backed up by a Cat sixteen-speed powershift and the cab was a top-shelf place to work.

▼ A four-wheel-drive articulated tractor, the MT975B was the world's most powerful tractor in early 2007.

JIM ALLEN COLLECTION

151 · Tech Backward

2008 New Holland TV6070 · Origin: **Fargo, North Dakota** · Company: **CNH Global**

By Jim Allen

▼ **The New Holland TV6070 was the last BiDi tractor, having debuted in 2008. It's shown with the optional lift, hitch, and PTO apparatus mounted at the "front," and the operator station facing that way.** NEW HOLLAND

New Holland continued the BiDi tradition with the TV6070, which was introduced in 2008. It used essentially the same chassis as the TV145 but with a new engine and many technology upgrades. The TV6070 was powered by a 6.7-liter New Holland six-cylinder, which is a bit smaller than the TV145's 7.5-liter engine but makes more power, delivers better fuel economy, and has lower emissions. The New Holland bidirectional tractor was discontinued, and the last one produced was built at the plant in Fargo, North Dakota, in December 2014.

152 · The Retro Movement

2009 New Holland Boomer 8N · Origin: **Matsumoto, Japan** · Company: **CNH Global**

By Jim Allen

With more than half a million original N-series Ford tractors built, New Holland looked to cash in on some history with a retro-styled tip of the hat to the original. In February 2009, New Holland introduced the Boomer 8N. Based on their successful Boomer line of compact tractors (introduced in 2006), the new 8N embodied the essence of the ancestral 8N, not only in the classic look but also in the practical collection of features. New Holland even tossed in a little style and glitz.

The Boomer 8N was manufactured for New Holland by Shibaura of Japan, who had been in the compact tractor business since 1950 and had a long history of building compact tractors for New Holland. It was powered by Shibaura's 2.2-liter (135-cubic-inch) ISM N844L four-cylinder diesel, a naturally aspirated four-stroke engine with indirect injection that made 50 flywheel horsepower at a fairly fast 2800 rpm.

What set the 8N apart mechanically from the rest of the Boomer line, even those with similar layouts and power outputs, was the CVT (Continuously Variable Transmission). It allowed infinite speed and gear adjustments from a pace where you could barely see the tires move to a top speed of 18.6 miles per hour.

The 8N was also a four-wheel-drive so it could put a big chunk of its 50 horsepower to the ground. To contrast, the original 8N, with a 28-horsepower, 119-cubic-inch flathead four-cylinder, was tested to a maximum of about 19 drawbar horsepower with a full load of ballast.

▲ **The Boomer 8N appeared in the original 8N red-and-gray color scheme and with styling that channeled the ancestor's retro look, while also looking remarkably modern and stylish.** NEW HOLLAND

153 · The Warrior

2013 Deutz-Fahr Agrotron 7250 TTV Warrior · Origin: **Lauingen, Germany** · Company: **SAME Deutz-Fahr**

▼ The special edition Warrior version of Deutz-Fahr's Agrotron 7250 TTV features jet-black bodywork in place of the usual light-green livery, and a CVT as standard.
SAME Deutz-Fahr

Special edition tractors have long been used to boost a brand's presence in particular market sectors, and although color has traditionally been a distinctive identifier of different brands, sometimes makers have repainted their machines to make them stand out. Deutz-Fahr's Warrior version of its Agrotron 7250 TTV is a prime example, with black paint replacing its usual green livery. With the TTV transmission, Deutz-Fahr was another CVT pioneer.

154 · Global Heritage

2013 Massey Ferguson 7624 · Origin: **Beauvais, France** · Company: **AGCO**

Massey Ferguson holds perhaps the most tortured brand history of any color, and also one of the strongest heritages of solid global sales. Stable for a decade or so in the AGCO family, the brand is doing the founders proud.

◀ **Built in France, the 7624 features a Sisu 449-cubic-inch turbo-diesel good for 210 PTO horsepower when tested at Nebraska in 2012.** AGCO

155 · Modern Power

2015 9620R · Origin: **Moline, Illinois** · Company: **John Deere**

High-horsepower tractors have evolved into incredibly complex machines, packed with a blend of technology, power, and sophistication. With price tags well above half a million dollars, you can think of them as a Veyron for the field and not be far off. The John Deere 9620R raised the bar for the big machines in 2015, and the 9620RX appeared in 2016 with four tracks.

◀ **The John Deere 9620R blends technology and horsepower in a distinctly American tractor package.** JOHN DEERE ARCHIVES

156 · Retro Mods

Big Bud 730/50 · Origin: **Havre, Montana** · Company: **Big Bud, LLC**

▶ **This Big Bud 730/50 is an upgraded and rebuilt machine.**

Marcus Pasveer

The appeal of high-horsepower machines with minimal electronics remains strong today, and a number of manufacturers exist just to rebuild and upgrade vintage machines to make them bullet-proof and utilitarian on modern farms. Big Bud, LLC and owner Ron Harmon do just that, and customers can have old machines rebuilt and made better than new. As of 2018, Harmon was also working on an all-new Series 5 Big Bud. The release date was not yet determined.

157 · European Refinement

2016 Steyr Terrus CVT · Origin: **St. Valentin, Austria** · Company: **Steyr Traktoren**

▶ **The Terrus CVT is powered by an Iveco 6.7-liter six-cylinder engine rated for 300 horsepower. It came to the United States as the Case IH Optum CVX line.**

Steyr Tracktoren

While American tractors rule the world in terms of sheer size, brute horsepower, and raw gadgetry, the European models pack their punch into sophisticated, high-speed machines able to work the ground and travel roads at high speed, plunge into muddy and at times mountainous conditions, and pamper the operator with sophisticated interiors.

158 · The Economical Beast

2016 Fendt 1050 Vario · Origin: **Marktoberdorf, Germany** · Company: **AGCO**

By Martin Rickatson

When introduced in 2016, Fendt's 517 engine horsepower 1050 Vario was the highest-horsepower conventional (FWA) tractor available worldwide. Like all Fendt tractors, it features CVT as standard, while further technology includes variable tire pressure and variable torque distribution between front and rear axles. In North America, the machine and its siblings are also sold in Challenger livery.

▼ Though Fendt does not have an articulated four-wheel-drive line, its 1000 Vario series tractors offer power outputs up to and just exceeding 500 horsepower, well into pivot-steer territory. AGCO

159 · Six Decades

2017 Steiger 620 Anniversary Edition · Origin: **Fargo, North Dakota** · Company: **Case IH**

Steiger celebrated sixty years of history in 2017, and Case IH commemorated that by releasing a limited number of its top-of-the-line 620 Quadtrac with this special paint scheme. The Steiger 620 Quadtrac was cutting-edge in 2017, combining high horsepower, a refined track system, and available auto-guidance and telemetry systems.

▲ **This 2017 Steiger 620 Anniversary Edition is shown with an original Quadtrac in a hanger at Chanute Air Force Base.**
Lee Klancher

160 · Back in the Saddle

2018 Versatile 610 · Origin: **Winnipeg, Canada** · Company: **Versatile**

In 1987, Canada's Versatile became part of Ford New Holland, and the four-wheel-drives were rebadged and decked out in blue livery. When Ford sold its ag interests to Fiat in 1991, Versatile was part of the deal. When Fiat then also purchased Case Corporation in 1999, it divested Versatile to meet anti-trust laws, as Case already owned Steiger. Canadian purchaser, businessman John Buhler, resurrected the Versatile name and colors before later selling a majority share to Rostselmash of Russia.

▶ **The top of the 2018 Versatile line, the 610 is available in wheeled and DT (Delta Track) versions, the latter fitted with rubber-track drive units on each corner. Its Cummins QSX15 engine produces a maximum 650 horsepower.**
Versatile

161 · High-Tech Agility

2018 Valtra N Series · Origin: **Suolahti, Finland** · Company: **AGCO**

Finnish tractor maker Valtra's N series tractors can be specified with an articulation point ahead of the cab, combined with conventionally steered front wheels, further improving maneuverability, and even allowing the tractor to be crab steered. The articulated steering and conventional front axle are automated, reducing the degree of articulation as speed increases. Studies suggest articulated tractors can perform front loader tasks 35–40 percent faster than standard tractors.

▶ **Valtra's N series tractors could be specified as standard rigid chassis models or with an articulation point ahead of the cab, combined with conventionally steered front wheels.** AGCO

Epilogue
The Future of the Tractor

2016 Case IH ACV · Origin: **Racine, Wisconsin** · Company: **Case IH**

At the 2016 Farm Progress Days, I remember sitting down with several veteran Case IH product managers who were blown away by the fact that they had just been in a session with a young man who had started driving tractors in 2007.

The man had never steered a tractor down a row, and in fact had no idea what that really meant.

That's nearly a decade of farm tractors being guided (down the rows, at least) by electronic systems rather than people, and a broad indicator of where the machines will go. Auto-guidance has long been something machines can do better than almost all operators, and the features have become a part of most any well-equipped modern machine.

Since the 1950s, the overwhelming trend has been to build larger tractors that can do more with one operator. This has been a response to the need to work more ground with less labor that has tormented the farmer for nearly seven decades.

The size and scale of the machines today is stunning, and some are so efficient that they work hard only a handful of days throughout the year. They can get the job done that fast.

An interesting issue I've heard is that when the autonomous tractor appears— and the technology is pretty much ready and in fact has been in the works since the early 1990s—tractors for the first time in ages could get smaller. A small autonomous tractor could work around the clock, and several of them would provide more flexibility and potentially economy for the farmer.

The autonomous concept has not been widely embraced by farmers, with a mixed reaction to the machine, so it may be quite some time before an all-new autonomous vehicle is released to the public. You can expect to see more features of that sort on the cutting-edge machines from each company.

Whatever comes in the days ahead, the technology evolution of the past 120 years has been an incredible ride. I can't wait to see what the future holds.

▼ **This is a Case IH concept of an autonomous tractor. It was first displayed to the public at the 2016 Farm Progress Show and created interest that rivaled or exceeded that of the original Quadtrac.** CASE IH

Bibliography

Allen, Jim. "1923 Advance-Rumely Oil Pull." *Diesel World*. February 23, 2017. https://www.dieselworldmag.com/features/early-oil-burner/.

American History. "Alkali Fuel Cells." Accessed July 1, 2018. http://americanhistory.si.edu/fuelcells/alk/alkmain.htm.

———. "Fuel Cell Basics." Accessed July 1, 2018. http://americanhistory.si.edu/fuelcells/basics.htm.

American Society of Mechanical Engineers Landmarks Program. "Hart-Parr Tractor." Accessed May 10, 2018. https://www.asme.org/about-asme/who-we-are/engineering-history/landmarks/190-hart-parr-tractor.

Baumheckel, Ralph. *International Harvester Farm Equipment Product History 1831-1985*. St. Joseph, MI: American Society of Agricultural and Biological Engineers, 1997.

Blackford, Mansel, and K. Austin Kerr. *Business Enterprise in American History*, 3rd ed. Belmont, CA: Wadsworth Publishing, 1993.

Broehl, Wayne. *John Deere's Company: A History of Deere and Company and Its Times*. New York City, NY: Doubleday, 1984.

Browne, William P., and John Dinse. "The Emergence of the American Agriculture Movement, 1977–1979." *Great Plains Quarterly* 5, no. 4 (1985): 221-35.

Clarke, Sally. "New Deal Regulation and the Revolution in American Farm Productivity: A Case Study of the Diffusion of the Tractor in the Corn Belt, 1920–1940." *Journal of Economic History* 51, no. 1 (March 1991): 101–123.

Culbertson, John. *The Tractor Builders: The People behind the Production of Hart-Parr/Oliver/White*. Charles City, IA: Sunrise Hill Associates, 2001.

Dahlstrom, Neil, and Jeremy Dahlstrom, *The John Deere Story: A Biography of Plowmakers John and Charles Deere*. Dekalb, IL: Northern Illinois University Press, 2005.

Deere & Company. "The Original Steel Plow," Accessed May 15, 2018. https://www.deere.com/en/our-company/history/john-deere-plow/.

Erb, David, and Eldon Brumbaugh. *Full Steam Ahead: J. I. Case Tractors and Equipment 1842–1955*. St. Joseph, MI: American Society off Agricultural and Biological Engineers, 1996.

Erie Art Museum. "Styled by Adams: Streamlining America, 1934–1958." Accessed May 19, 2018. https://erieartmuseum.org/styled-by-adams-streamlining-america-1934-1958/.

Fairbanks Morse. "Fairbanks Morse History." Accessed May 10, 2018, http://www.fairbanksmorse.com/about/history/.

Garvey, Scott. "How Engineers Invented the Rubber-Belted Ag Tractor." Grainews.

November 7, 2014. https://www.grainews.ca/2014/11/07/how-engineers-invented-the-rubber-belted-tractor-for-ag/.

Gas Engine Magazine. "The Massey Harris 4 Wheel Drive." December 1978. https://www.gasenginemagazine.com/tractors/the-massey-harris-4-wheel-drive.

Gray, R. B., *The Agricultural Tractor, 1855–1950*. St. Joseph, MI: American Society of Agricultural and Biological Engineers, 1980.

Green Magazine. "John Deere Custom Color Lawn Tractors." December 12, 2017. https://greenmagazine.com/custom-color-lawn-tractors/.

———. "The John Deere 4020 New Generation Tractor: An Instant Classic." January 2, 2018. https://greenmagazine.com/john-deere-4020/.

Haycraft, William. *Yellow Steel*. Champaign, IL: University of Illinois Press, 2000.

The Henry Ford. "Ford Motor Co. Chronology." Accessed May 5, 2018. https://www.thehenryford.com/collections-and-research/digital-resources/popular-topics/ford-company-chronology/.

The History Museum South Bend. "The Oliver Chilled Plow Works." Accessed May 20, 2018. https://historymuseumsb.org/the-oliver-chilled-plow-works/.

Hyams, Joe. "James Dean." *The Independent*. September 27, 2015. https://www.independent.co.uk/news/people/james-dean-press-photographer-dennis-stock-on-how-he-came-to-know-and-love-the-doomed-screen-idol-a6669611.html.

Jacobson, David. "Founding Fathers." *Stanford Magazine*. May 2, 2018.

The John Deere Journal. "Model 'M' Tractor Was New from the Ground Up." April 8, 2016. https://johndeerejournal.com/2016/04/the-model-m-innovation-economy-feature/.

Kapko, Matt. "History of Apple and Microsoft: 4 Decades of Peaks and Valleys." *CIO*. October 5, 2015.

Klancher, Lee. *The Art of the John Deere Tractor*. Austin, TX: Octane Press, 2011.

———. *The Farmall Dynasty: A History of International Harvester Tractors*. Austin, TX: Octane Press, 2008.

———. *Red Tractors, 1958–2018*. Austin, TX: Octane Press, 2018.

———. "The Solar Aircraft Company and the HT-340." April 18, 2017. https://octanepress.com/internationalharvesterHT340.

Klancher, Lee, and Jim Allen. *Red 4WD Tractors*. Austin, TX: Octane Press, 2017.

Levingston, Steve. "As the Auto Age Dawned, Gasoline Wasn't King." *Washington Post*. August 13, 2006. http://www.washingtonpost.com/wpdyn/content/article/2006/08/12/AR2006081200214.html.

Macmillan, Don. *The John Deere Tractor Legacy*. North Humberside, UK: Japonica Press, 2003.

McCormick, Sargent M., and Barry Machado. "Irony in American Business History." *Built in Chicago*. October 20, 2013. https://www.builtinchicago.org/blog/chicago-entrepreneurial-history-mccormick-bothers-and-rise-megacorps.

Miller, Merle L. *Designing the New Generation John Deere Tractors*. St. Joseph, MI: American Society of Agricultural and Biological Engineers, 1999.

Morrell, Herbert. *Oliver Tractors*. 2nd ed. Austin, TX: Octane Press, 2012.

Newman, Jesse, and Jacob Bunge. "The Transformation of the American Farm, in 18 Charts." *Wall Street Journal*. December 28, 2017. https://www.wsj.com/articles/the-transformation-of-the-american-farm-in-18-charts-1514474480?ns=prod/accounts-wsj.

Newsweek Special Edition. "The Story of Steve Jobs, Xerox and Who Really Invented the Personal Computer." *Newsweek*. March 19, 2016. https://www.newsweek.com/silicon-valley-apple-steve-jobs-xerox-437972.

O'Clair, Jim. "1948 Allis-Chalmers Model G." *Hemmings Motor News*. November 2014.

Peterson, Greg. "The 5 Millionth IHC Tractor 1974 Model 1066 with 53 Hours." AgWeb. May 25, 2016. https://www.agweb.com/agday/blog/machinery-pete/the-5-millionth-ihc-tractor-1974-model-1066-with-53-hours/.

Porsche-Diesel North American Registry. "History." Accessed July 2, 2018. http://www.porsche-diesel.com/history.aspx.

Preuhs, Dave. "New Hart-Parr." Hart-Parr Tractors. 2002. http://hartparrtractors.tripod.com/newhartparr.html.

Ryan, Camille, and Jamie M. Lewis. "Computer and Internet Use in the United States: 2015.' *American Community Survey Reports*. September 2017. United States Census Bureau. https://www.census.gov/content/dam/Census/library/publications/2017/acs/acs-37.pdf.

Schaefer, Sherry. *Classic Oliver Tractors*. Minneapolis, MN: Voyageur Press, 2014.

Seyfarth, A. C., *Tractor History, 1910–1934*. Madison, WI: Historical Society Press, 2008.

Simpson, Peter. *The Big Bud Tractor Story*. Greenville, SC: 3-Point Ink, 2017.

———. *Ultimate Tractor Power Vol. 2, M–Z: Articulated Tractors of the World*. North Humberside, UK: Japonica Press, 2002.

Smithsonian Institution. "Smithsonian Online Virtual Archives." Accessed July 1, 2018. https://sova.si.edu/.

The Smithsonian National Museum of History. "Hart-Parr #3 Tractor." Accessed

May 18, 2018. http://americanhistory.si.edu/collections/search/object/nmah_857022.

Snider, David. "The Last Word on James Dean and Dennis Stock." *New York Times*. February 6, 2016. https://lens.blogs.nytimes.com/2016/02/08/james-dean-dennis-stock-life-lens-photos/.

Spielmaker, D. M. "Growing a Nation Historical Timeline." Ag Classroom. March 21, 2018. https://www.agclassroom.org/gan/timeline/farmers_land.htm

Strohl, Daniel. "1938 Farm On! – 1938 Minneapolis Moline UDLX Comfortractor from Hemmings Classic Car." Accessed May 17, 2018. https://www.hemmings.com/magazine/hcc/2005/04/1938-Farm-On----1938-Minneapolis-Moline-UDLX-Comfortractor/1280675.html.

Swinford, Norm. *Allis-Chalmers Farm Equipment 1914–1985*. St. Joseph, MI: American Society of Agricultural and Biological Engineers, 1994.

———. *A Century of Ford and New Holland Farm Equipment*. Washington, DC: American Society of Association Executives, 2000.

Swinford, Norm. *The Proud Heritage of AGCO Tractors*. St. Joseph, MI: American Society of Agricultural and Biological Engineers, 1999.

Thompson, Derek, "How the Tractor (Yes, the Tractor) Explains the Middle Class Crisis." *Atlantic*, March 13, 2012. https://www.theatlantic.com/business/archive/2012/03/how-the-tractor-yes-the-tractor-explains-the-middle-class-crisis/254270.

TractorData. "Custom C." Accessed May 19, 2018. http://www.tractordata.com/farm-tractors/005/4/2/5421-custom-c.html.

"Tractormobile Diesels for General Purpose Farm Work." *Implement and Tractor*. May 22, 1945. http://www.wagnertractors.com/documents/Wagnerannouncement22May1954.pdf.

Troyer, Howard. *The Four Wheel Drive Story*. New York City, NY: McGraw-Hill Company, 1954.

United States v. J. I. Case Co., 101 F. Supp. 856 (D. Minn. 1951).

Wilkerson, Isabel. "The Man Who Put Steam in Your Iron." *New York Times*, July 11, 1991. https://www.nytimes.com/1991/07/11/garden/the-man-who-put-steam-in-your-iron.html.

Will, Oscar H., III. "The 1206: Making the Most of the DR-361." April 27, 2017. https://octanepress.com/content/1206.

Will, Oscar H., III. *Garden Tractors: Deere, Cub Cadet, Wheel Horse, and All the Rest, 1930s to Current*. Minneapolis, MN: Voyageur Press, 2009.

Acknowledgments

Every book is a team effort. I was privileged to work with this talented crew to make this book come to life.

The book creation team of designer Tom Heffron, project editor Aki Neumann, copyeditor Chelsey Clammer, and proofer Dana Henricks, provided vital attention to detail, vision, and all the critical guidance provided by creative eyes.

Authors and photographers Jim Allen, Scott Garvey, Larry Gay, Michael Hood, Marcus Pasveer, Martin Rickatson, Sherry Schaefer, and Peter Simpson offered their text, images, and expertise.

Our team of fact-checkers and advance readers included Dave Clausen of Old Abe News, Cheryl Delap of Prairie Gold Rush, Dave Mowitz of Successful Farming, and collectors Lou Buice and Chris Wathen.

Thanks to the former executives and engineers who helped out, including Rich Hale, Chuck Pelly, George Russell, George Vater, and Bud Youle.

Thanks to collectors who took the time to get their machines out. Special thanks to Max Armstrong, Lou and J. T. Buice, Ben Colburn, Glenn Glass, the late Darius Harms, Jim Johnson, Bruce Keller, Jerry Kuster, Brian Look, Pete McCann, Jerry Mez, Jack Purinton, Doug Rumboldt, Vernon Smith, Jerry Smoker, Mary Wakeman, Bill Weir, Geoff Wing, and Coy Winslow.

Thanks to Joan Hughes, my wife, for her insight, guidance, and support.

▼ **Henry and Marion Van Ruler have more than six hundred tractors in Western Minnesota, with almost all of them in their original "work clothes." Cheers to the collectors!**
Lee Klancher

Index